BROOKINGS INSTITUTION, WASHINGTON, D.C.
INSTITUTE FOR GOVERNMENT RESEARCH

SERVICE MONOGRAPHS
OF THE
UNITED STATES GOVERNMENT
No. 1

THE U. S. GEOLOGICAL SURVEY

ITS HISTORY, ACTIVITIES AND ORGANIZATION

AMS PRESS
NEW YORK

THE INSTITUTE FOR GOVERNMENT RESEARCH

Washington, D. C.

The Institute for Government Research is an association of citizens for coöperating with public officials in the scientific study of administrative methods with a view to promoting efficiency in government and advancing the science of administration. It aims to bring into existence such information and materials as will aid in the formation of public opinion, and will assist officials, particularly those of the national government, in their efforts to put the public administration upon a more efficient basis.

To this end, it seeks by the thoroughgoing study and examination of the best administrative practice, public and private, American and foreign, to formulate those principles which lie at the basis of all sound administration, and to determine their proper adaptation to the specific needs of our public administration.

The accomplishment of specific reforms the Institute recognizes to be the task of those who are charged with the responsibility of legislation and administration; but it seeks to assist, by scientific study and research, in laying a solid foundation of information and experience upon which such reforms may be successfully built.

While some of the Institute's studies find application only in the form of practical coöperation with the administrative officers directly concerned, many are of interest to other administrators and of general educational value. The results of such studies the Institute purposes to publish in such form as will insure for them the widest possible utilization.

INSTITUTE FOR GOVERNMENT RESEARCH

SERVICE MONOGRAPHS
OF THE
UNITED STATES GOVERNMENT
No. 1

THE U. S. GEOLOGICAL SURVEY

ITS HISTORY, ACTIVITIES AND ORGANIZATION

D. APPLETON AND COMPANY
NEW YORK LONDON
1918

Library of Congress Cataloging in Publication Data

Brookings Institution, Washington, D.C. Institute for Government Research.
The U. S. Geological Survey.

Original ed. issued as no. 1 of Service monographs of the United States Government.
Bibliography: p.
1. United States. Geological Survey. I. Series: Brookings Institution, Washington, D.C. Institute for Government Research. Service monographs of the United States Government, no. 1.
QE76.B7 1974 353'.008'232 72-3014
ISBN 0-404-57101-8

Reprinted from the edition of 1918, New York and London
First AMS edition published, 1974
Manufactured in the United States of America

International Standard Book Number:
Complete Set: 0-404-57100-X
Volume 1: 0-404-57101-8

AMS Press, Inc.
New York, N.Y. 10003

PUBLICATIONS OF THE INSTITUTE FOR GOVERNMENT RESEARCH

STUDIES IN ADMINISTRATION

The System of Financial Administration of Great Britain
By W. F. Willoughby, W. W. Willoughby and S. M. Lindsay

The Budget
By René Stourm
T. Plazinski, Translator, W. F. McCaleb, Editor

The Canadian Budgetary System
By H. G. Villard and W. W. Willoughby

The Problem of a National Budget
By W. F. Willoughby

The Movement for Budgetary Reform in the States
By W. F. Willoughby

Teachers' Pension Systems in the United States
By Paul Studensky

Organized Efforts for the Improvement of Methods of Administration in the United States
By Gustavus A. Weber

The System of Financial Administration of the United States

PRINCIPLES OF ADMINISTRATION

Principles Governing the Retirement of Public Employees
By Lewis Meriam

Principles of Government Purchasing
By A. G. Thomas

SERVICE MONOGRAPHS OF THE UNITED STATES GOVERNMENT

The United States Geological Survey

The Reclamation Service

FOREWORD

The first essential to efficient administration of any enterprise is full knowledge of its present make-up and operation. Without full and complete information before them, as to existing organization, personnel, plant, and methods of operation and control, neither legislators nor administrators can properly perform their functions.

The greater the work, the more varied the activities engaged in, and the more complex the organization employed, the more imperative becomes the necessity that this information shall be available—and available in such a form that it can readily be utilized.

Of all undertakings, none in the United States, and few, if any, in the world, approach in magnitude, complexity, and importance that of the national government of the United States. As President Taft expressed it in his message to Congress of January 17, 1912, in referring to the inquiry being made under his direction into the efficiency and economy of the methods of prosecuting public business, the activities of the national government "are almost as varied as those of the entire business world. The operations of the government affect the interest of every person living within the jurisdiction of the United States. Its organization embraces stations and centers of work located in every city and in many local subdivisions of the country. Its gross expenditures amount to nearly $1,000,000,000 annually. Including the personnel of the military and naval establishments, more than 400,000 persons are required to do the work imposed by law upon the executive branch of the government.

"This vast organization has never been studied in detail as one piece of administrative mechanism. Never have the foundations been laid for a thorough consideration of the relations of all of its parts. No comprehensive effort has been made to list its multifarious activities or to group them in such a way as to present a clear picture of what the government is doing. Never has a complete description been given of the agencies through which these activities are performed. At no time has the attempt been made to study all of these activities and agencies with a view to the assignment of each activity to the agency best fitted for its performance, to the avoidance of duplication of plant and work, to the integration of all administrative agencies of the government, so far as may be practicable, into a unified organization for the most effective and economical dispatch of public business."

To lay the basis for such a comprehensive study of the organization and operations of the national government as President Taft outlines, the Institute for Government Research has undertaken the preparation of a series of monographs, of which the present study is one, giving a detailed description of each of the fifty or more distinct services of the government. These studies are being vigorously prosecuted, and it is hoped that all services of the government will be covered in a comparatively brief space of time. Thereafter, revisions of the monographs will be made from time to time as need arises, to the end that they may, as far as is practicable, represent current conditions.

These monographs are all prepared according to a uniform plan. They give: first, the history of the establishment and development of the service; second, its functions, described not in general terms, but by detailing its specific activities; third, its organization for the handling of these activities; fourth, the character of its plant; fifth, a compilation of, or reference to, the laws and regulations governing its operations; sixth, financial statements showing its appropriations, expenditures and other data for a period of years; and finally, a full bibliography of the sources of information, official and private, bearing on the service and its operations.

In the preparation of these monographs the Institute has kept steadily in mind the aim to produce documents that will be of direct value and assistance in the administration of public affairs. To executive officials they offer valuable tools of administration. Through them, such officers can, with a minimum of effort, inform themselves regarding the details, not only of their own services, but of others with whose facilities, activities and methods it is desirable that they should be familiar. Under present conditions services frequently engage in activities in ignorance of the fact that the work projected has already been done, or is in process of execution by other services. Many cases exist where one service could make effective use of the organization, plant or results of other services had they knowledge that such facilities were in existence. With the constant shifting of directing personnel that takes place in the administrative branch of the national government, the existence of means by which incoming officials may thus readily secure information regarding their own and other services is a matter of great importance.

To members of Congress the monograph should prove of no less value. At present these officials are called upon to legislate and appropriate money for services concerning whose needs and real problems they can secure but imperfect information. That the possession by each member of a set of monographs, such as is here projected, prepared according to a uniform plan, will be a great aid to intelligent legislation and appropriation of funds can hardly be questioned.

To the public, finally, these monographs will give that knowledge of the organization and operations of their government which must be had if an enlightened public opinion is to be brought to bear upon the conduct of governmental affairs.

These studies are wholly descriptive in character. No attempt is made in them to subject the conditions described to criticism, nor to indicate features in respect to which changes might with advantage be made. Upon administrators themselves falls responsibility for making or proposing changes which will result in the improvement of methods of administration. The primary aim of outside agencies should be to emphasize this responsibility and facilitate its fulfillment.

While the monographs thus make no direct recommendations for improvement, they cannot fail greatly to stimulate efforts in that direction. Prepared as they are according to a uniform plan, and setting forth as they do the activities, plant, organization, personnel and laws governing the several services of the government, they will automatically, as it were, reveal, for example, the extent to which work in the same field is being performed by different services, and thus furnish the information that is essential to a consideration of the great question of the better distribution and coördination of activities among the several departments, establishments and bureaus, and the elimination of duplications of plant, organization and work. Through them it will also be possible to subject any particular feature of the administrative work of the government to exhaustive study, to determine, for example, what facilities, in the way of laboratories and other plant and equipment, exist for the prosecution of any line of work and where those facilities are located; or what work is being done in any field of administration or research, such as the promotion, protection and regulation of the maritime interests of the country, the planning and execution of works of an engineering character, or the collection, compilation and publication of statistical data, or what differences of practice prevail in respect to organization, classification, appointment, and promotion of personnel.

To recapitulate, the monographs will serve the double purpose of furnishing an essential tool for efficient legislation, administration and popular control, and of laying the basis for critical and constructive work on the part of those upon whom responsibility for such work primarily rests.

Whenever possible the language of official statements or reports has been employed, and it has not been practicable in all cases to make specific indication of the language so quoted.

CONTENTS

THE U. S. GEOLOGICAL SURVEY: ITS HISTORY, ACTIVITIES AND ORGANIZATION

CHAPTER I

HISTORY

The United States Geological Survey, a bureau of the Department of the Interior, is engaged chiefly in surveying the geology, topography, and related features of the United States and in publishing the results of its surveys and investigations.

Early Explorations and Surveys of the Public Domain: 1804-1865. Previous to the establishment of the Geological Survey, the surveys conducted by the national government were chiefly exploratory in character, and were, therefore, confined almost wholly to the western country. Numerous surveys were made at an early date, most of them by the Army, partly for military use and partly to extend geographical knowledge. The most important was the Lewis and Clark Expedition of 1804-1806, which ascended the Missouri to its sources and then descended the Columbia to the Pacific. The expeditions of Pike in 1805 and 1807 to the sources of the Mississippi and the Arkansas were also of prime importance. In 1820 an important expedition in Upper Michigan, Wisconsin and Minnesota was conducted by General Cass, then Superintendent of Indian Affairs; and in 1832 an expedition to one of the sources of the Mississippi was made by Schoolcraft, the famous student of Indian life, while traveling in behalf of the Bureau of Indian Affairs.

While as early as 1820 General Cass had taken a mineralogist with him, it was not until 1834 that Congress made a specific appropriation for geological surveys. The act of June 28, 1834, appropriated $5,000 "to be applied to geological and mineralogical survey and researches." This appropriation was used in making a geological survey of the country between the Missouri River and the Red River, and the resulting report contained, in addition to an account of the geology and mineralogy of the areas covered, a geologic section of the country from the New Jersey coast to the Red River in Texas.

In the period from 1835 to 1850 the surveys and explorations made in the western country under Army auspices rapidly increased in number. Of these perhaps the most noteworthy were those of Captain, later General Fremont. Of the civilian surveys of the period, those made by David Dale Owen, beginning in 1839, under the auspices of the General Land Office, were of great importance, embracing as they did the geology of the region now included in the states of Illinois, Iowa, Wisconsin and Minnesota.

During 1853-1861 the War Department, under authority of the act of March 3, 1853 (10 Stat. L., 219) and supplementary acts, made surveys for a railroad from the Mississippi River to the Pacific Ocean, geologists accompanying each of the field parties. The investigations made were reported in thirteen quarto volumes, published in 1855-1860 under the title, "Reports of explorations and surveys to ascertain the most practicable and economical route for a railroad from the Mississippi River to the Pacific Ocean."

During the Civil War and for a year or more thereafter no government survey of note appears to have been undertaken.

Geographical and Geological Surveys: 1867-1879. With the close of the war the work of surveying and mapping the western country was renewed with increased vigor. Between 1867 and 1869 four surveying expeditions were put in the field, two by the War Department and two by the Department of the Interior, as follows:

1. *Geological Exploration of the Fortieth Parallel.* The act of March 2, 1867 (14 Stat. L., 457) provided for "a geological and topographical exploration of the territory between the Rocky Mountains and the Sierra Nevada Mountains, including the route or routes of the Pacific Railroad." This survey was made under the jurisdiction of the War Department but its director, Clarence King, was a civilian, as were also all his scientific assistants. The territory surveyed comprised a belt 105 miles in width extending from longitude 104° 30′ to longitude 120°,—that is, from Cheyenne, Wyoming, to the eastern boundary of California. The survey was primarily geological in character but included also the topography of the region. The results of the survey were published in 1870-1880 in seven volumes and an atlas. The total cost of the survey and its publications was $383,711.

2. *Geological and Geographical Survey of the Territories.* The act of March 2, 1867 (14 Stat. L., 471, sec. 2) called for "a geological survey of Nebraska, said survey to be prosecuted under the direction of the Commissioner of the General Land Office." F. V. Hayden was assigned to this work and was subsequently designated United States geologist for the territories of Colorado and New Mexico. As the work progressed, its scope was by authority of Congress extended over all the territories and work was done in Wyoming, Idaho, Montana, New Mexico and Colorado, the total area embraced in systematic surveys reaching about 100,000 square miles. This survey was primarily geological, but its activities included work on topography, geology, paleontology, ethnology, philology, botany and kindred sciences. The results were published in a series of volumes which were issued from 1867 to 1883.[1]

[1] The titles of the reports for 1867-1872 do not contain the word "geographical;" those for 1873-1878 do contain that word. The titles of the bulletins read: Bulletin of the United States Geological and Geographical Survey of the Territories.

The original appropriation for this survey was only $5,000, but this was increased by successive annual appropriations of $10,000, $25,000, $50,000, $75,000, $115,000, $110,000, $95,000, $65,000, $75,000 and $75,000, as well as by a specific appropriation of $30,000 to complete maps and office work; so that the total cost of the survey amounted to $735,000 exclusive of the cost of printing and engraving and of the services of several officers detailed from the Army.

3. *Geographical and Geological Survey of the Rocky Mountain Region.* In 1867 the Smithsonian Institution began an exploration of the Colorado River. This survey was later recognized by Congress in a joint resolution, approved July 11, 1868 (15 Stat. L., 253), authorizing the Secretary of War "to issue rations for twenty-five men of the expedition engaged in the exploration of the river Colorado under direction of Professor Powell, while engaged in that work." Additional appropriations, $54,000 in all, were granted in 1870-73, the expedition still remaining under the control of the Smithsonian Institution. On the completion of the survey of the Colorado River, Powell was, by act of June 23, 1874 (18 Stat. L., 207), authorized to continue the survey in Utah under the direction of the Secretary of the Interior; and subsequent acts of appropriation authorized the extension of the field of survey to the "Rocky Mountain region." In all, the area surveyed was 67,000 square miles, embracing southern Wyoming, central and southern Utah and adjacent portions of Nevada and Arizona. This survey was primarily exploratory and geographical but, in addition to the triangulation of the whole region and the establishment of the geodetic points, it included work in topography, ethnology, geology, botany, paleontology and kindred sciences.

The results of the survey were not published in full, the only printed documents produced being two brief reports in 1877 and 1878 by Powell.

Though the appropriations for the survey amounted only

to $244,000, the total cost, not including engraving and printing, as stated by Powell in a letter to the Secretary of the Interior, was $279,000.

4. *Geographical Survey West of the One Hundredth Meridian.* An act approved June 10, 1872 (17 Stat. L., 367) authorized a "continuance of military and geographical surveys and explorations west of the one hundredth meridian of longitude." The survey was made under the jurisdiction of the War Department, Lieutenant George M. Wheeler, of the Engineer Corps, being placed in charge. This survey, as its name implies, covered all the territory west of the one hundredth meridian, which includes the western parts of the Dakotas, Nebraska, Kansas and Texas, the Rocky Mountain states, and the Pacific Coast states. This survey was primarily geographical or topographical, but, as was stated in the report of the Chief of Engineers to the Secretary of War on May 10, 1878 (House Executive Document No. 88, 45th Congress, 2d Session), was so made as to obtain "at the same time and as far as practicable without greatly increasing the cost, all the information necessary before the settlement of the country, concerning the branches of mineralogy and mining, geology, paleontology, zoology, botany, archæology, ethnology, philology, and ruins."

The survey was brought to a close in 1879 and its results were published in 1875-1889 under varying titles.[2]

The original appropriation for this survey was $75,000, but a series of additional appropriations for continuing the survey, for engraving and printing, and for completing the office work of the survey brought the total direct appropriations for the survey up to the sum of $618,644. If to this sum is added the value of the aid and supplies received from the War Department and the salaries of regular Army officers detailed

[2] The titles of these reports and of the papers they contain are given in Bulletin 222 of the U. S. Geological Survey.

to the survey, the cost of the survey appears to have been $805,340.[3]

Establishment of the U. S. Geological Survey, 1879. It will thus be seen that in the early seventies four surveys were simultaneously in progress under specific appropriations made by Congress, two under the Interior Department and two under the War Department, each having a method of its own and in several places two covering the same area.

In addition, the Coast and Geodetic Survey, after having executed a triangulation and detailed survey of part of the coast of the United States, had, under authority granted by Congress in the act of March 3, 1871, extended its work into the interior in order to provide, by means of primary triangulation carried across the continent, geodetic connection between the Atlantic and Pacific Coasts.

The feeling became very strong, both in Congress and outside, that this condition of affairs was unsatisfactory, for there was evidently not only loss of efficiency and unnecessary expense by having five separate organizations working in the same general field, but duplication and overlap of work.[4]

As a result, in 1878, Congress in making further appropriations for the existing surveys (act of June 20, 1878, 20 Stat. L., 230) provided that:

The National Academy of Sciences is hereby required, at their next meeting, to take into consideration the methods and expenses of conducting all surveys of a scientific character under the War or Interior Department, and the surveys of the Land Office, and to report to Congress as soon thereafter as may be practicable a plan for surveying and mapping the Territories of the United States on such general system as will, in their judgment, secure the best results at the least possible cost; and also to recommend to Congress a suitable plan for

[3] A full discussion of the indirect cost of this survey may be found in the report of the commission to consider the organization of the Signal Service, Geological Survey, etc., 1886. (Senate Report No. 1285, 49th Congress, 1st Session.)

[4] See H. R. Report No. 612, 43d Congress, 1st Session.

the publication and distribution of the reports, maps, and documents, and other results of said surveys.

The National Academy submitted its report to the House of Representatives on December 3, 1878 (House Executive Document No. 5, 45th Congress, 3d Session). In this report the academy, after briefly discussing the systems of survey then existing, recommended that the Coast and Geodetic Survey be transferred from the Treasury to the Interior Department and there take over all surveys of mensuration, including those being performed by the four surveys heretofore enumerated, as well as the work performed by the surveyors of the General Land Office. The report proceeds to state that:

> The best interests of the public domain require, for the purposes of intelligent administration, a thorough knowledge of its geological structure, natural resources, and products. The domain embraces a vast mineral wealth in its soils, metals, salines, stones, clays, etc. To meet the requirements of existing laws in the disposition of the agricultural, mineral, pastoral, timber, desert, and swamp lands, a thorough investigation and classification of the acreage of the public domain is imperatively demanded. The committee, therefore, recommend that Congress establish, under the Department of the Interior, an independent organization, to be known as the United States Geological Survey, to be charged with the study of the geological structure and economical resources of the public domain, such survey to be placed under a Director, who shall be appointed by the President, and who shall report directly to the Secretary of the Interior.

The recommendations of this report were incorporated in the legislative, executive and judicial appropriation bill, and in this form favorably acted upon by the House, but in the Senate, owing to the congestion of business at the close of the session, and for other reasons not connected with the merits of the matter, these provisions failed of passage. In the resulting conference committee the provisions intended to consolidate all surveys of mensuration in a single service to be organized around the Coast and Geodetic Survey were dropped, but the following was agreed on for insertion in the sundry civil appropriation bill:

For the salary of the Director of the Geological Survey, which office is hereby established, under the Interior Department, who shall be appointed by the President, by and with the advice and consent of the Senate, $6,000; provided, that this officer shall have the direction of the Geological Survey,[5] and the classification of the public lands and examination of the Geological Structure,[5] mineral resources, and products of the national domain, and that the Director and members of the Geological Survey shall have no personal or private interests in the lands or mineral wealth of the region under survey, and shall execute no surveys or examinations for private parties or corporations; and the geological and geographical survey of the Territories, and the geographical and geological survey of the Rocky Mountain region, under the Department of the Interior, and the geographical surveys west of the one hundredth meridian, under the War Department, are hereby discontinued, to take effect on the 30th day of June, 1879. And all collections of rocks, minerals, soils, fossils, and objects of natural history, archæology, and ethnology, made by the Coast and Interior Survey, the Geological Survey, or by any other parties for the Government of the United States, when no longer needed for investigations in progress, shall be deposited in the National Museum.

For the expenses of the Geological Survey and the classification of the public lands and examination of the geological structure, mineral resources and products of the national domain, to be expended under the direction of the Secretary of the Interior, $100,000. * * *

The publications of the Geological Survey shall consist of the annual report of operations, geological and economic maps illustrating the resources and classification of the lands, and reports upon general and economic geology and paleontology. The annual report of operations of the Geological Survey shall accompany the annual report of the Secretary of the Interior. All special memoirs and reports of said Survey shall be issued

[5] The capitalization of these words, which appears in the Statutes (Vol. 20, p. 394), is doubtless a typographical error.

in uniform quarto series if deemed necessary by the Director, but otherwise in ordinary octavos.

On March 3, 1879, the bill containing these provisions became law (20 Stat. L., 394).

Pursuant to the provisions of this law, the President appointed as Director of the Geological Survey, Clarence King, who, as already stated, had directed the geological exploration of the fortieth parallel. Mr. King organized a corps of geologists, most of whom had been connected with one or another of the former government surveys, among them being Major John W. Powell, who had conducted the exploration of the Colorado River and the geographical and geological survey of the Rocky Mountain region. This corps was organized by Mr. King in four sections, each to operate, under the direction of a chief, in a definite geographical district. For this purpose the territory west of the one hundred and second meridian was divided into four districts.

Extension of Operations to the Country at Large. As has been seen, the law of 1879 creating the Geological Survey authorized it to operate in the "national domain." This phrase was interpreted by Director King as referring only to "the region of the public land." At the same time, however, he urged upon the congressional committee the desirability of extending the operations of the Survey over the country as a whole. On June 28, 1879, the House of Representatives accordingly passed a joint resolution amending the act of March 3, 1879, by inserting after the words "national domain," the words, "and he may extend his examination into the states." In the Senate this joint resolution was reported back from committee with amendments authorizing such extension with the consent of the states, but it did not come to a vote.

The case for operations on a national scale was forcefully argued by the Director in his first annual report.[6]

[6] P. 77 ff.

The great extent of the United States and the widely separated sources of the national resources render the acquisition by private citizens of information on almost any single product always difficult, often impossible.

As a direct result of the size of the country, the government and people have long been uninformed as to our primary industries; those, I mean, which yield the raw materials—mineral, vegetable, and animal.

To the Agricultural Department we owe the first reforms from this condition of wide-spread ignorance. In the realm of mineral productions the only efforts made to acquire any positive knowledge have been the highly useful, but feebly endowed, works of the late mining commissions, whose investigations were suffered to end for lack of appropriations.

Today no one knows, with the slightest approach to accuracy, the status of the mineral industry, either technically, as regards the progress and development making in methods, or statistically, as regards the sources, amounts, and valuations of the various productions.

Statesmen and economists, in whose hands rest the subjects of tariff and taxation, have no better sources of information than the guesses of newspapers and the scarcely less responsible estimates of officials who possess no adequate means of arriving at truth.

In no other intelligent nation is this so; on the contrary, mineral production is studied with the most elaborate effort. England, France, Germany, Austria, Russia, and Italy consider it essential to know, from year to year, not only the source and aggregates of amount and value of mineral yield, but many lesser facts relating to the modes and economies of the industries.

Upon considering the extent of country over which our minerals occur, their wonderful variety and yet unmeasured amounts, it cannot fail to be apparent that no private individual or power is competent to do what ought long since to have been done, namely, to sustain a thoroughly practical investigation and exposition of the mineral industry.

By way of example, and to show how hopeless it is to look to any other source than the government for this service, I select iron.

The lack of comprehensive knowledge of the occurrence and distribution of iron, despite numerous state and private investigations, is then set forth at length:

> To claim that the iron question will ever be adequately investigated as a whole, either by private enterprise or State surveys, is to betray a total lack of appreciation of the character, magnitude, and needs of the industry.
>
> What is true of this single metal is equally true of nearly the whole catalogue of the mineral products of the United States. A few exceptional items, like quicksilver, occur in such restricted areas that private or State enterprises could contribute all the knowable facts and features of the business of production. But as a whole it is true, and can never be refuted, that the Federal Government alone can successfully prosecute the noble work of investigating and making known the natural mineral wealth of the country, current modes of mining and metallurgy, and the industrial statistics of production.
>
> Provided Congress extends the field of the Geological Survey over the whole national territory, and appropriates the comparatively small amounts necessary for the maintenance of the organization, it will be entirely practicable to carry forward this work, and contribute powerful aid to the mineral industries. Of the desirableness, from every point of view, of the results of a general geological survey, I conceive there cannot be two opinions. That these results can only be attained by an organization under Federal patronage, is, in my opinion, scientifically certain.

It was not until two years later that the recommendation of the Director was adopted by Congress and then only in part. The sundry civil appropriation act of 1882 (act of August 7, 1882, 22 Stat. L., 329) added to the original words of appropriation, which had been repeated in the act of 1880 and 1881, the words, "and to continue the preparation of a geologic map of the United States." It will be observed that the provision for "examination of the mineral resources and products" which had been made in the original act with respect to the national domain was not by this act extended to the country at large. It was not indeed until 1888 that all restrictions confining the work of the Survey to the public land region disappeared from the appropriation acts. In that year appropriation was made "for geologic surveys in various portions of the United States," a clause since repeated annually. Under this clause, which is much broader than the former clause authorizing the preparation of a geologic map, the Survey has extended every type of geologic investigation to the country at large.

As recently as 1914 the question was raised in Congress whether the work of the Survey should not be limited, either absolutely or conditionally, to the public lands, but no proposals embodying this view have ever received favorable consideration in either House.

In 1884 an agreement for coöperation was made between the Geological Survey and the Commonwealth of Massachusetts, which was about to undertake a topographic survey of the state, the expense of field work to be divided equally between the national survey and the state, the national survey to engrave the maps and give transfers of the plates to the state. This agreement, which worked out very satisfactorily, was but the first of a long series of agreements with the surveys of a number of the states for coöperation in topographic and subsequently in hydrographic and geologic work, and this form of state coöperation has been of the greatest value in permitting the extension of the Survey's standard methods and operations over the country at large much more rapidly than might otherwise have been possible. Coöperative agreements in geologic and paleontologic work were in effect with twenty-five states on June 30, 1917.[7]

As early as 1889 the Survey had informally extended its operations to the territory of Alaska, having sent its repre-

[7] No attempt is made here to describe this system of coöperation in full, as the Survey has described it in detail in a pamphlet entitled "Coöperation between the United States and the Various States in Topographic, Hydrographic and Geologic Work, 1910." The extent to which states have coöperated up to June 30, 1917, is, however, shown in the report of the Director of the Survey for the fiscal year 1917, and is summarized in Appendix 5.

sentatives with various governmental and private exploring expeditions with a view to gathering geological data. It first became possible, however, to undertake comprehensive work in Alaska in 1895, when Congress appropriated $5,000 for the investigation of the gold and coal resources of Alaska, and this amount was again appropriated in 1896 and 1897. Early in 1898, after rich gold deposits had been found in the Klondike region, in Canadian territory, the appropriation was increased to $25,000. In 1901 the scope of the surveys in Alaska was extended to cover all mineral resources and the annual appropriation was increased to $60,000. In 1905 it was increased to $80,000, in 1909 to $90,000, and in 1912 and later years to $100,000. A total of over one and a half million dollars has thus been expended by the Survey on Alaskan surveys and investigations. Before the Survey entered Alaska only the most meager information was available not only as to its resources but even as to the major features of its geography. In thirty years the Survey has mapped topographically over one-third of its surface and has made geologic explorations, in greater or less detail, over almost the whole of this part of its area.

Even prior to their acquisition by the United States, the Survey had from time to time made geologic researches in Hawaii, Cuba and Porto Rico. In 1899 Congress by resolution called upon the Survey for information as to existing surveys of our insular possessions, and for estimates of the cost of further surveys. No action was taken by Congress on the report presented, however, nor has any express authority ever been granted to the Survey to extend its operations to those possessions. In 1910, however, the Survey entered into a cooperative agreement with the territorial government of Hawaii for the topographic survey of the territory which had already been begun by the territorial government, and work under similar agreements has since been steadily prosecuted. Similarly, in 1912, an agreement was entered into by the Survey with the Isthmian Canal Commission and the Smithsonian Institution for geologic studies in the Canal Zone.

Development of Survey's Functions. The broad purposes of the Survey's work have remained constant throughout its history, but its functions and the manner of their performance have undergone many changes from time to time. These will here be briefly outlined, detailed discussion of the development of particular activities being reserved for the following chapter.

The organic act of 1879, together with the amending act of 1882, specified four distinct functions to be performed by the Survey—the geological survey of the public lands, the classification of the public lands, the examination of the geological structure and the mineral resources and products of the national domain, and the preparation of the geologic map of the United States.

At first the Survey took the position that by the term "classification of the public lands," Congress did not intend to imply a detailed survey of the individual tracts of land in the public domain, with respect to their inclusion under the provisions of one or another of the public land laws, but merely a general survey of the soil and mineral characteristics of the lands. In effect, therefore, this portion of the law was regarded as adding little if anything to the clauses calling for the geological survey of the public lands and for the examination of the mineral resources and products of the national domain, and in this interpretation Congress apparently acquiesced, as it was followed by the Survey for many years without evoking any change in the phraseology of the successive appropriation acts.

The authority to prepare a geologic map of the country was construed by the Survey as giving authority to prepare also a topographic map; the position being taken that a geologic map could properly be prepared only if a topographic map were available as a base.

In accordance with the interpretation of the statutes the early work of the Survey fell into three classes: (1) topographic surveys; (2) geologic surveys in areas of economic importance or theoretical interest; (3) research in geologic theory and the theory of related sciences.

In 1882 the scope of the Survey's functions was enlarged by a provision in the appropriation act for the year 1882-3 that "not to exceed ten thousand dollars of the amount appropriated may be applied under the direction of the Secretary of the Interior to the procuring of statistics in relation to mines and mining other than gold and silver and in making chemical analysis of iron, coal and oil." Almost immediately upon its organization the Survey had undertaken to collect statistics of mines and mining in coöperation with the Tenth Census. The legislation quoted placed the work upon a permanent basis and permitted the publication by the Survey in 1883 of a volume entitled "Mineral Resources of the United States," comprising over 800 pages of statistical matter for the calendar year 1882. An appropriation for the preparation of this report has since been granted every year, and the report has come to be one of the best known of the Survey's publications.

Congressional Investigation of 1886. The Survey had been in existence but little over five years when its plans and methods became the subject of a sweeping inquiry by a joint congressional commission. The rapidly growing appropriations for several large scientific bureaus of the government—the Geological Survey, the Signal Service (which then performed the functions since vested in the Weather Bureau), and the Hydrographic Office—became, beginning in 1883, a subject of discussion in Congress, and in 1884 provision was made in the sundry civil appropriation act (23 Stat. L., 219) for a joint congressional commission to "consider the present organizations" of the bureaus mentioned, "with the view to secure greater efficiency and economy in the administration of the public service in said Bureaus." The commission appointed held numerous hearings in 1885 and 1886, Major

Powell, the Director, appearing for the Survey.[8] The main questions regarding the Survey's work raised in hearings related to the value of topographic maps executed on so small a scale as was then employed by the Survey, the rate of progress toward completion and publication of the Survey's geological maps, and the propriety of the wide scope of the Survey's scientific investigations and publications.

While the minority of the commission, consisting of two of the six members, submitted a report criticizing the Survey on the score of costliness, slowness and lack of practical applicability of its work, the majority expressed the opinion that the Survey as a whole was "well conducted, and with economy and care, and discloses excellent administrative and business ability on the part of its chief." [9] They reported, however, in favor of restricting the printing or engraving of "the scientific reports known as Monographs and Bulletins of the Geological Survey" by requiring that specific and detailed estimates should be made therefor and specific appropriations made in pursuance of such estimates. This provision was accordingly incorporated in the sundry civil appropriation act adopted shortly afterward (act of August 4, 1886, 24 Stat. L., 255).

In 1888 the appropriation for the operating expenses of the Survey, which had up that time been granted as a lump sum, was itemized under eight heads; and of a total appropriation of slightly over $737,240, $40,000 was appropriated for "paleontologic researches" and $17,000 for "chemical and physical researches." Thus, within two years after the commission had made its report, Congress by express action sustained the Survey in its paleontologic and chemical and physical research, the utility of which had been questioned by

[8] Major Powell's testimony furnishes a mine of information relative to the early operations of the Survey. All the testimony taken by the commission was printed as Senate Misc. Doc. No. 82, 49th Congress, 1st Session (Vol. 4 of Misc. Docs.).

[9] Senate Report No. 1285, 49th Congress, 1st Session.

the minority of the commission, and these provisions became permanent features of the appropriation acts of the Survey.

The Powell Irrigation Survey, 1888-1890. The Director of the Survey, Major Powell, had for some years been deeply interested in the question of reclamation of the arid public lands by irrigation. In 1878 he had rendered to the Commissioner of the General Land Office a "report on the land of the arid region of the United States with a more detailed account of the land of Utah." In discussing the methods to be employed for the reclamation of the arid lands, he had stated that it involved engineering problems requiring for their solution the greatest skill, as well as the employment of large capital and possibly government aid, if not direct government construction. By 1888 the continued appropriation and reduction to private ownership of the readily available streams in the arid region, which he had foreseen, resulted in the authorization by Congress of investigation by the Geological Survey of "that portion of the arid regions of the United States where agriculture is carried on by means of irrigation, as to the natural advantages for the storage of water for irrigation purposes, with the practicability of constructing reservoirs, together with the capacity of the streams and the cost of construction and capacity of reservoirs, and such other facts as bear on the question of storage of water for irrigating purposes." By an appropriation act passed a few months subsequently and by one passed in 1889, a total of $350,000 was appropriated for the conduct of this investigation.

The act of October 2, 1888 (25 Stat. L., 526), however, in addition to making appropriation for the investigation, directed the Survey to designate all lands which might be used "for sites for reservoirs, ditches, or canals for irrigation purposes, and all the lands made susceptible of irrigation by such reservoirs, ditches, or canals;" and it was provided that all such lands "are from this time henceforth hereby reserved from sale as the property of the United States, and shall not be subject after the passage of this act to entry, settlement or occupation until further provided by law." Thus, for the first time, the Survey was vested with powers of direct administration with respect to the public domain.

Immediately upon the enactment of this legislation, the Survey organized a large force, composed principally of topographic engineers, and undertook a vigorous prosecution of the project entrusted to it. Within a little more than a year after its organization this branch of the Survey, which is commonly referred to as the Powell Irrigation Survey, segregated 127 reservoir sites, having an area of over 2,500 square miles, and, in addition, over 30,000,000 acres of irrigable land located in five distinct basins.[10]

The sweeping action of the Survey provoked a wide-spread protest from persons in the arid region who were adversely affected and as a result a special committee of the Senate was appointed to investigate the entire subject. The report of the majority of the committee, rendered in 1890, severely criticized the policy adopted by the Survey.[11] By the act of August 30, 1890 (26 Stat. L., 391), the whole of the act of 1888 was accordingly repealed except as to reservoir sites, the segregation and reservation of which was expressly continued. The appropriations made in 1888 and 1889 for the survey of the arid lands were also discontinued.

Though it may be said that the operations of the Powell Irrigation Survey thus met with congressional disapproval, the actual results achieved by it in topographic maps of the arid regions and in stream measurements constituted a work of enduring value and furnished the basis upon which the subsequent work of the Survey in connection with reclamation was largely based.

Whether because of the disfavor with which its operations

[10] The reports on the operations of the Powell Irrigation Survey were published in separate volumes of the annual report of the Director of the Survey for the fiscal years 1889 to 1892, inclusive. The figures are taken from the volume for 1889, page viii.

[11] The report of the committee together with the record of the hearings held by it was printed as Senate Report No. 928, 51st Congress, 1st Session.

in the arid lands had been viewed by Congress or because of the general tendency to reduce appropriations as a result of the economic depression in the country, the appropriations to the Survey suffered a heavy reduction in the period from 1890 to 1896. Up to and including the fiscal year ending June 30, 1890, the Survey's appropriations had increased with every year, the increase being in one year as high as $150,000, so that for the fiscal year 1890 the total amount appropriated was over $875,000. For the fiscal year 1891 the appropriation was reduced to less than $850,000, but the reduction was due entirely to the elimination of the irrigation survey, the appropriations for topographic and geologic surveys receiving a marked increase. In the following year the appropriation for topographic surveys was somewhat reduced, so that the total appropriations were but little more than $700,000. During the next three years, however, the reductions were exceedingly heavy, the amount for topographic surveys being reduced successively from $250,000 to $240,000, $200,000 and $150,000; but the appropriation for geologic surveys, at first reduced from $115,000 to $50,000, was subsequently increased to $70,000 and $100,000. Proportionately, the heaviest reductions were made in the relatively small appropriations for paleontological researches and chemical and physical researches, the former being reduced from $40,000 in 1892 to $10,000 for the three following years, and the latter from $17,000 in 1892 to $5,000 in 1893 and 1894, and $7,000 in 1895. In 1895, accordingly, the total appropriation for the Survey was but slightly over $500,000; less than had been appropriated for the year 1886. With the fiscal year 1896 a period of substantial increase again set in which has continued without material interruption until the present time.[12]

The figures presented show that in the early nineties the topographic work of the Survey had become perhaps its major activity as determined by percentage of cost. The grow-

[12] All the appropriations for the Survey are shown in detail in tabular form in Appendix 5.

ing importance of this work was further emphasized by the third director of the Survey, Charles D. Walcott, who succeeded Major Powell in 1894. Mr. Walcott, in his first annual report, announced his intention of enhancing the value of the topographic survey for other than geologic purposes and ordered the representation upon the topographic maps of land subdivision lines and township and section corners in the public land states, and he obtained authority from Congress to print and sell the topographic maps with text for educational purposes.

The desirability of having detailed topographic maps of an area in advance of its geologic or other special investigation resulted, very early in the Survey's history, in the extension of topographic work over large areas that had not been covered by geologic investigations. So extensive and accurate was the topographic work thus performed that it rapidly assumed an importance of its own quite independent of its prospective utilization in connection with the Survey's geologic or other work. In 1888 this condition was recognized by Congress by a specific appropriation for the first time for topographic work, and in the next year by the transfer to the Geological Survey of the work of engraving the topographic and geologic maps—work that had formerly been entrusted to private engravers, under contract with the Government Printing Office.

In 1896, by a special act of Congress, a new function was temporarily imposed upon the Survey, that of conducting a land subdivision survey. The work was done in the Indian Territory during the years 1896 to 1898. The Director of the Survey reported that this work had demonstrated "that it is more economical to survey large areas in this manner than under the contract system heretofore employed by the Government in its land subdivision surveys. This statement applies to large areas embracing one thousand square miles or more." With respect to smaller areas the Director was of the opinion that the system of employing local surveyors under contract was the more economical. No subsequent land subdivision surveys have been required of the Survey by Congress. In 1898 there was, however, an appropriation made for the survey by the Geological Survey of the northern portion of the Idaho-Montana boundary, and from time to time other boundary surveys have been made by the Survey.

The history of the Survey's geologic and topographic work in the next decade is one of steady progress along established lines, and a rapid development of that part of its work which had direct economic application. The extension of its operations to Alaska, beginning in 1895, has already been mentioned. During these years, in addition to thus entering a field where rapid economic development was imminent, the Survey attacked in succession three major problems of the conservation of the natural resources of the country—the conservation of the national forest lands, the reclamation of the arid public lands, and the elimination of waste, both of raw material and of life, in the mining industries of the country.

Survey of Forest Reserves. From its beginning, in connection with its regular geologic and topographic surveys, the Survey had gathered data relative to the country's forests; and in 1891, on the enactment of a law authorizing the President to set aside forest reserves on the public lands, the Survey had been able to give advice to the Secretary of the Interior as to the determination of the boundaries of such reserves. The lack of definite information as to the conditions and resources of the lands included in the forest reserves had made impracticable, however, any attempt to administer them. In 1897, therefore, Congress appropriated $150,000 "for the survey of the public lands that have been or may hereafter be designated as forest reserves by Executive proclamation," specifying at the same time that these surveys should be made under the supervision of the Director of the Survey (30 Stat. L., 34). Under this authority the Survey began a thorough survey of the national forests—their distribution, the size and density of the timber, the distribution of the leading economic species, the damage inflicted by fires, the amount of dead timber, the extent to which the forests are pastured, the amount of timber already cut and the effects of deforesting, the relations of timber supply and transportation, the local demands of miners and settlers, and the supply needed for more distant markets. Upon the basis of the information thus gathered, regulations for the use of the forest reserves were framed and the boundaries of the reserves were readjusted.

The examinations of the forest reserves were carried steadily forward for seven or eight years and covered about 75,000,000 acres. In addition to reports on the several reserves, the Survey undertook, in connection with these examinations, the preparation of land classification maps of the areas examined. These maps were prepared on the regular atlas sheets of the Survey as bases, and showed not only the forests with burnt and cut lands but the irrigable and pasture lands. In all some forty sheets of this type were prepared and published.

During the years 1897 to 1905 the data collected by the Survey relative to the forest reserves furnished the basis of the regulations governing those reserves and of the administration of those regulations. The actual administration of the reserves was, however, vested in the General Land Office. In 1905 the administration was transferred to the Bureau of Forestry of the Department of Agriculture, which was renamed the Forest Service. To this Service was also transferred all further examination and classification of the forests; and the work of the Survey relating to forests was henceforth limited to "topographic surveys on public lands which have been or may hereafter be designated as national forests," an activity for which specific appropriations, varying in amount from $100,000 to $75,000, have since annually been made.

Stream Gaging and Reclamation Work. After the repeal, in 1890, of those parts of the act of 1888 that provided for the segregation of sites for canals and ditches and of lands made available for irrigation, the appropriations made to the Survey during 1888 and 1889 for surveys in the arid lands had also been discontinued. In 1894 the Survey, however, obtained a specific appropriation of $12,500 "for gauging the streams and determining the water supply of the United States, including the investigations of underground currents and artesian wells in arid and semiarid sections." This small appropriation was doubled for the fiscal year 1896 and that amount again doubled for the fiscal year 1897, so that for the three years 1897 to 1899 it stood at $50,000, rising to $70,000 in 1900 and $100,000 in 1901, the terms of the appropriation being broadened, however, in 1896 to read: " * * * for gauging streams and determining the water supply of the United States, the investigation of underground currents and artesian wells and the preparation of reports upon the best methods of utilizing the water resources." As a result of the operations conducted under these appropriations, which, though they extended over the whole country, were naturally concentrated in the arid regions, the Survey accumulated a most extensive body of information regarding the water resources of the arid regions, particularly with reference to their availability for irrigation. It was but natural, therefore, that upon the passage, in 1902, of the reclamation act, the administration of the act should be entrusted by the Secretary of the Interior to the Survey.

The Reclamation Service was first organized merely as a division of the so-called Hydrographic Branch, the chief of that branch acting also as chief engineer of the Reclamation Service. As the work of that service passed, however, from the stage of planning to that of actual construction, its association with the Hydrographic Branch and then with the Survey itself became more and more nominal until, in 1907, pursuant to a recommendation made by the Director of the Survey, it became an independent service, subject to control only by the Secretary of the Interior, its chief engineer, the former chief of the Hydrographic Branch of the Survey, becoming its director. Since that time the relation between the Survey and the Reclamation Service has been merely that of coöperation in respect to certain matters of stream measurement.

Work in Mining Technology. From the first the Survey had established a close relation to the mining and mineral industries of the country through the preparation of its annual report on "Mineral Resources of the United States." This relation was steadily strengthened by investigations made by the Survey in important mining areas, by investigations and publications relative to technologic processes, and by the publication, beginning in 1894, of parts of a geologic map of the country, of which the early issues covered, almost without exception, areas of interest to the mining industry. In 1898 the growth of this relation of the Survey to the mining industry was reflected in the introduction into Congress of a resolution [13] calling for the creation, by statute, in the Survey, of a separate division of mines and mining, with a specific appropriation, on the ground that the mining interests of the country should have "a clearly defined representation in the organization of the Government." Its passage was recommended by the Survey, but without success.

Some six years later Congress authorized the Survey to undertake a kind of work which had been mentioned by the Director, in his report recommending the establishment of a division of mines and mining,[14] as one which might well be taken up by such a division—a systematic inquiry into the values of the several deposits of economic minerals in the country. By act of February 18, 1904, there was appropriated the sum of $30,000 "for analyzing and testing the coals and lignites of the United States." In 1905 this appropriation was enlarged to $227,000 and was extended to cover all fuels; and simultaneously there was added an appropriation for investigating structural materials. In the next few years the appropriations for both these purposes were largely increased.

In connection with both investigations, as they proceeded,

[13] Senate Res. No. 205, 55th Congress, 3d Session.
[14] Annual Report for 1898-1899, p. 21.

the Survey was called upon to give expert advice to other branches of the government as to the fuels and structural materials used by them; and the preparation of specifications and the testing of fuels and materials came to be a standing feature of the Survey's work.

The emphasis which had thus tended, in this field of the Survey's work, toward the technologic rather than the geologic aspect of the mineral industries was strengthened by an appropriation made in 1908 (act of May 22, 1908) "for the protection of lives of miners in the territories and in the districts of Alaska, and for conducting investigations as to the causes of mine explosions with a view to increasing safety in mining." Under this appropriation the Survey made examinations of explosives used in coal mining in the United States, of the occurrence of explosive gases and inflammable or explosive dust, of safety lamps and mine-rescue apparatus, and of the use of electricity in mines. These investigations, like the forestry surveys, necessitated the addition to the Survey's personnel of an entirely new staff of technical experts—mining technologists.

For the fiscal year 1908-09 the appropriations for the three classes of technologic investigations just described—testing of fuels, testing of structural materials, and investigation of mine explosives—had reached a total of over $500,000, and though in the next fiscal year the item for testing fuels was reduced from $250,000 to $100,000, it was evident that the investigations of mining accidents would continue to grow in cost and importance. The lack of intimate connection between these investigations and the regular geologic work of the Survey, and the feeling, which, as already seen, had been expressed in Congress more than ten years before, that the mining industries of the country should have definite representation in the government, resulted in a proposal that a Bureau of Mines be created as an independent bureau of the Department of the Interior, and that it take over and develop the work in mining technology begun by the Survey. By act of May 16, 1910, this proposal was enacted into law, and on July 1 the personnel of the Technologic Branch of the Survey was organized as the Bureau of Mines, the chief of that branch becoming the first director of the new bureau.

By the same act, the investigation of structural materials was transferred to the Bureau of Standards.

Land Classification Work. The three important phases of the Survey's work just traced—in forestry, reclamation, and mining technology—had all been undertaken in pursuance of special statutory authority. About 1905, however, an activity of prime importance was begun without any change in the law.

The very first clause in the definition of functions in the act of 1879 creating the Survey had provided that the Survey should have charge of "the classification of the public lands." This phrase, however, was from the first interpreted by the Survey, and inferentially also by Congress, as calling for a mere geologic and mineralogic examination of the public lands without reference to the public land laws. For over twenty years this interpretation governed, but then the General Land Office and the Secretary of the Interior began to call more frequently upon the Survey for its opinion, based on information either already gathered by it or obtained by special investigation, as to the applicability of particular provisions of the land laws to specific tracts of public land to which patent was sought by private parties or with respect to which official action was contemplated. In 1906 the Survey undertook, with the approval of the Secretary of the Interior, a systematic valuation of the public lands that were believed to contain coal, in order to determine the prices at which such lands should be sold by the government under the coal land law. This work is still in progress and to it has been added the examination and classification of oil, phosphate, potash, and mineral lands, and of lands suitable for water-power sites, as well as enlarged homesteads and stock-raising homesteads. Today the classification of the public lands by the Survey constitutes an integral part of their administration.

In the last decade the functions of the Survey have not materially changed. Rather there has been a marked development of the existing functions of the Survey in the direction of increasing the popularity and practical value of the results of its work. This tendency is seen particularly in the more popular character of some of the publications, in the more prompt publication of statistics of mineral resources, and in the vigorous prosecution of geologic and mineralogic inquiries having for their purpose the location of deposits of phosphates, nitrates, petroleum, potash, and other minerals for which there is an urgent industrial need.

The Survey and the War. Many of the early surveys of the West were largely military. Among their objects was the ascertainment of the most practicable routes of transportation in a country not supplied with railroads, the topography of the country as bearing on the conduct of military operations, and the resources and development of the country with respect to its ability to furnish supplies for troops. As the West developed and as the Geological Survey extended its topographic mapping over larger and larger areas of the country, the need for separate military surveys largely disappeared and the maintenance of a corps of topographical engineers by the Army was discontinued.

In 1916 the War Department received an appropriation of $35,000 for the conduct of certain topographical surveys deemed necessary in connection with the development of plans for the national defense. An agreement was entered into between the War Department and the Survey by which the appropriation was expended by the Survey under the direction of the War Department. In the execution of this work the Survey prepared a specimen topographical map similar to that developed by the best European practice in military mapping.

In March, 1917, the recognition of the urgent needs of the

military service resulted in the plans for the Survey's entire field work in topography being made to conform to a program drawn up by the General Staff of the Army. The map printing plant of the Survey was also made available to the War and Navy Departments for which it has printed a great number of charts and maps.

In addition to these military topographic surveys, the Survey made special examinations of underground waters and soil drainage in a number of localities that were under consideration by the War Department for military encampments.

Of no less importance for the prosecution of the war are the special investigations which have been made by the Survey with a view to locating new deposits of minerals, both metals and non-metals, required in the manufacture of munitions. Among these minerals are manganese, pyrite, platinum, chromite, tungsten, antimony, potash, quicksilver, and nitrate.

Perhaps the most striking contribution made by the Survey to the military service has been the group of specially trained engineers in its personnel who have been commissioned in the Engineer Officers' Reserve Corps. At the end of the fiscal year 1917, this group included one geologist, seven hydrographic engineers, and sixty-one topographic engineers.

CHAPTER II

ACTIVITIES

In the preceding chapter the development of the activities of the Survey has been briefly outlined. In the present chapter each of these activities will be treated in turn and the attempt will be made to trace in some detail its origin and development, its present purpose, the method of its performance and the progress made to date in accomplishing the program proposed.

It may be said that all the activities of the Survey have, broadly speaking, two purposes—that of making public information regarding the topography and resources of the country, and that of obtaining information necessary for the proper administration of the laws governing the disposition of the public lands.

As indicated in the preceding chapter, though the general character of the Survey's activities is determined by the permanent provisions of laws establishing and governing the Survey, which provide that the Director of the Survey shall have charge of "the geological survey and the classification of the public lands and examination of the geological structure, mineral resources and products of the national domain, . . . shall continue the preparation of a geologic map of the United States," yet the extent and the specific nature of the work may be determined by the annual appropriations made for the Survey's work. The appropriation acts for the year ending June 30, 1918, specifically authorize expenditures for the following activities:

1. Topographic surveys in various portions of the United States.

2. Geologic surveys in various portions of the United States.

3. Chemical and physical researches relating to the geology of the United States, including researches with a view of determining geological conditions favorable to the presence of deposits of potash salts.

4. Investigation of the mineral resources of Alaska.

5. Preparation of a report on the mineral resources of the United States.

6. Gaging streams and determining the water supply of the United States, the investigation of underground currents and artesian wells, and the preparation of reports on the best methods of utilizing the water resources.

7. Discovering, developing, protecting and rendering more accessible springs, streams and water holes on arid public lands of the United States; erecting and maintaining monuments and sign boards and providing ready and convenient means by which water may be brought to the earth's surface.

8. Examination and classification of lands requisite to and suitable for enlarged homesteads, stock-raising homesteads, public watering places and stock driveways as required by the public land laws.

The scope of the activities thus authorized is of course dependent upon the amount of the appropriation. As will hereafter be noted, the act further controls the scope of some specific activities by a special provision regarding the manner of the utilization of the amount appropriated.

For a clearer presentation the activities of the Survey may be regarded as falling under the following seven heads:

1. Geologic surveys and investigations.

2. Compilation of information relative to the mineral industries.

3. Topographic surveys.

4. Surveys and investigations relative to water resources.

5. Examinations, under act of March 1, 1911, of land proposed to be purchased by the government for the protection of navigable streams.

6. Investigations of the character and value of public lands made necessary by the land laws.

7. Operation of lithographing and engraving plant.

As the purpose of the larger part of the Survey's work is the acquisition and dissemination of knowledge concerning geography, geology, mineral resources, and water resources, most of its activities, therefore, involve the publication of the results achieved.

Topographic Survey. As already indicated, the Survey is engaged in making and publishing a topographic atlas of the United States—that is, a series of maps exhibiting the configuration of the land and water, the elevations of the land, and the principal features of culture. In addition to these atlas sheets it is publishing special maps of certain areas—such as the areas including large cities.

The work on the topographic atlas was begun by the Survey in 1882, promptly upon the enactment of the provision of law authorizing the Survey to prepare geologic maps of the United States. In entering upon the task, one of unprecedented magnitude, of executing a topographic survey of the entire United States, the Survey fell heir to the results which had been produced by previous topographic surveys, federal, state and private. Notable among these results were the maps produced by the several geographical and geological explorations and surveys of the West, to which reference was made in the preceding chapter. The materials thus acquired were of very unequal value, but they furnished a point of departure for the Survey's work.

The Survey did not for some years attempt to publish topographic maps for general distribution, but supplied them only to its own members and to other government services. Authority to sell the maps to the general public was finally granted by Congress in 1897 (act of February 18, 1897, 29 Stat. L., 701).

During the first few years of the Survey's map work it continued to use the same scale that had been employed by the earlier independent surveys—about 4 miles to the inch, or 1:250,000, with relief shown by 200-foot contours; but when the coöperative surveys of the states were begun it was seen to be desirable to increase the scale and decrease the contour interval, not only for the work in those states, but for that in other states in which coöperation was not being carried on.

The earlier topographic mapping of the Survey was subjected to some criticism on account of lack of refinement in detail. It was a question of judgment whether the appropriation should be expended in accurately mapping a limited area or in making reconnaissance surveys over a more extended area. The less accurate maps were less expensive, and served a useful purpose in establishing the value of topographic surveys. The results have proved the wisdom of a policy whereby the people were gradually educated to the use of refined and detailed maps, while Congress learned to appreciate the cost and the comparative slowness with which accurate surveys would be prosecuted over so vast a territory as that of the United States. Had fewer square miles been mapped annually in the early years, but with the accuracy attained at present, it is doubtful if Congress would have continued to maintain the organization. As it was, valuable results were rapidly put forth, and in time it has been possible to supersede the older reconnaissance maps with more detailed maps.[1]

As the topographic work steadily advanced, not only in the area covered, but in accuracy, it was found desirable to obtain from Congress authority to run careful spirit-level lines and to establish bench marks for permanent record of eleva-

[1] U. S. Geological Survey, Bulletin No. 227, 1904, p. 57.

tions upon the ground, and also to carry on control by triangulation or traverse with greater accuracy and to mark the positions more permanently than had been possible without statutory authority. Accordingly, on June 11, 1896, an act was approved which provided that "elevations above a base level located in each area under survey shall be determined and marked on the ground by iron or stone posts or bench marks," etc. Thereafter it was possible to provide for the running of careful spirit levels over all the areas under survey and to extend the amount of spirit leveling, thereby securing more accurate location of contours. Since that act was passed the topographic mapping has reached as high a degree of detail and quality as seems desirable for the scales adopted.

The selection of areas for survey is determined by a variety of considerations. It is of course desired to extend the benefits of the work as widely and equally over the country as a whole as is economically practicable. Other things being equal, however, it is more economical to extend surveys outward continuously from areas already surveyed than to undertake new work in areas entirely unconnected with them by triangulation or otherwise. A special economic need or project in an isolated area may nevertheless occasionally warrant the initiation of a survey there. As between areas equally economical of survey, preference is determined by a number of factors, among which are the apparent local demand for surveys, and the probable economic value of the resultant maps.

The methods of field work followed by the Survey in making topographic surveys are unique and have been developed by the topographers of the Survey through nearly forty years of field experience. Wherever possible the results of triangulation by other organizations are utilized. The degree of refinement in the detailed surveying depends upon the scale upon which the resulting map is to be drawn.

The results of the topographical surveys are published in the form of Topographic Atlas Sheets, measuring 17 by 20 inches. These sheets are published on different scales, the scale selected varying with the character of the area mapped and the degree of detail required. These scales are 1:62,500, 1:125,000, and 1:250,000; that is, approximately 1 mile, 2 miles, and 4 miles to the inch, respectively. A sheet thus represents an area of approximately 230 square miles, 920 square miles, or 3,700 square miles. The largest scale, showing the most detail and representing the smallest area, is used for densely populated or otherwise important areas; the next largest for sparsely inhabited mountain regions; the smallest for areas covered by reconnaissance or exploratory surveys, especially in the western states and Alaska.

The boundaries of each quadrangle, as the area mapped on a single sheet is known, are meridians of longitude and parallels of latitude, the differences in longitude or latitude between the boundaries being in the largest scale maps 15′, and in the smaller scales, 30′ and 1°, respectively.

The maps show the topographic features of the land, such as mountains, hills, valleys, and gulches; all bodies of water, such as lakes, marshes, streams, and springs; the routes of travel, such as railroads, wagon roads, and trails; political boundaries; cities, towns, and permanent buildings; and the names of natural and other features. They also indicate exactly the location of permanent survey monuments and bench marks whose positions and whose elevations above sea level have been determined by precise methods. The maps of areas covered by public land surveys show all township and section lines as well as the boundaries of all land grants.

Each of these maps is printed in three colors—black for the cultural features, such as boundary lines, roads, railroads, houses, towns, and cities, and for the names of all things represented; brown for the relief, or the element of elevation, indicated by contour lines showing the heights of all parts of the area above sea level; and blue for the water—the rivers, lakes, and the sea and its bays and inlets.

Each sheet is named from some prominent natural or other feature within the quadrangle it represents, such as a lake, mountain, town or city. Index maps showing the location and names of quadrangles mapped in any state or region are published.

In addition to regular topographic atlas sheets many special topographic maps have been published of areas having particular interest, such as large cities and their environs, or regions embracing natural wonders. The Survey has also been engaged for some time in participating in the preparation of the United States portion of the millionth scale map of the world—that is, a map having a scale of one to one million, or a scale of about sixteen miles to the inch. This map is to be part of an international map of the world upon a uniform scale. A few sheets of this map have already been published by the Survey. A number of base maps of the several states and of the United States are also published by the Survey.

Historically considered, the topographic maps produced by the Survey are primarily of use to the Survey itself as necessary bases for geologic maps, but they are now to be regarded as the basic and standard topographic maps of the country for all purposes. They are used by virtually all departments of the national, state, and local governments in the location of field offices, the determination of the boundaries of administrative districts and the consideration of projects for public works, a use in which they are of special value. Before these maps were made, for example, every city was obliged to spend large sums in surveys for water supply. These topographic maps now furnish all necessary data for a far larger area than would be embraced in a local survey. In engineering works, whether municipal or constructed by private capital, and in industrial, mining, and agricultural operations of every sort, the Survey's topographic maps are also of the highest value. The maps are widely used, moreover, in traveling and in countless other activities of every-day life. The extent of their use is indicated by the annual sales, which now amount to over half a million copies.

Attention should be called to the especial value which attaches to the Survey's topographic maps of Alaska, because of the undeveloped state of the country. Demands for them come from prospectors, engineers, capitalists, and school teachers, and they form one of the most essential preliminaries to any form of development of the country. They indicate routes of travel to the prospector and explorer, railway and wagon routes to the locating engineer, and possible sources of water power to the mine operator.

Cost and Progress. The cost of the surveys first made, on the scale of 4 miles to the inch, averaged $1.75 a square mile. Those made later, on the scale of about 2 miles to the inch, averaged in cost $4 a square mile, and those made on the scale of approximately 1 mile to the inch, averaged in cost $10 a square mile. During the field season of 1884 a single party mapped over 11,000 square miles on the scale of 4 miles to the inch. A few years later the output of a single party, on the scale of 2 miles to the inch, was about 3,000 square miles in a season. Today the more refined and detailed maps, on the scale of about 2 miles to the inch, with a contour interval of 100 feet, cost from $7 to $11 a square mile, according to the country, and a single party can rarely map over 600 to 1,000 square miles in a season. On the scale of 1 mile to the inch, a party rarely maps more than 500 square miles in a season, and the cost of this work varies between $12 and $30 a square mile, according to the nature of the country.

It would be difficult to compute with accuracy the total cost of the topographic surveys and of the topographic maps to date. As already stated, the Survey inherited from the several western surveys which preceded it a large body of material which it was able to utilize in the preparation of its early topographic maps. Again, it was not until eight years after the Survey began topographic surveys that specific appropriation was made for them by Congress. The state contributions to topographic work, amounting to nearly two and one-half million dollars, must also be taken into account. The total specific appropriations for topographic surveys to date have been over eight million dollars. Thus, it seems safe to say that the total cost of these surveys and of the preparation of the topographic maps has been not less than twelve million dollars.

Of the 3,026,789 square miles in the United States proper, 1,257,623 square miles have now been mapped, covering about 2,225 sheets. The accompanying map, taken from the annual report of the Director of the Survey for 1917, shows the progress that had been made up to June 30 of that year in the execution of the topographic survey of the country.

In addition, there have been mapped 202,361 square miles in Alaska, comprising over 34 per cent. of the total area; and 1,393 square miles in Hawaii, comprising 22 per cent. of the total area.

Erection of Bench Marks. In connection with its topographic work, the Survey erects bench marks in areas surveyed, a bench mark being a monument upon which is noted its elevation. This work has been carried on since 1896, in pursuance of an act of Congress passed in that year which required that thereafter in topographic surveys "west of the ninety-fifth meridian, elevations above a base-level located in each area under survey shall be determined and marked on the ground by iron or stone posts or permanent bench marks, at least two such posts or bench marks to be established in each township or equivalent area, except in the forest-clad and mountain areas, where at least one shall be established, and these shall be placed, whenever practicable, near the township corners of the public land surveys; and in the areas east of the ninety-fifth meridian at least one such post or bench mark shall be similarly established in each area equivalent to the area of a township of the public land survey."

A bench mark serves two purposes—that ordinarily served by a monument of establishing an exact point of reference in the use of the topographic maps (the location of the bench mark being indicated on the map), and that of furnishing for all purposes a permanent record on the land itself of its elevation. Bench marks also are of value in the progress of the Survey's topographic work as fixed points of reference for the extension of its level lines.

It will be noted that the act of 1896 does not require that the elevations of the bench marks above sea level be determined, but only "the elevations above a base level located in each area." Any attempt to determine elevations above sea level in the initial survey of an area might have necessitated the running of thousands of miles of precise levels in order to connect with sea level the initial points within each of the areas under topographic survey. This would have cost immense sums and would have occupied several years, during which the topographic surveys would have had to be continued without spirit leveling. The Survey, therefore, designated some point within each area under topographic survey as a central datum point for that area, and determined the elevation of that point as nearly as practicable from existing elevations adjusted by railway levels or other levels brought from the sea. In consequence, though all the elevations referred to the same central datum point agree one with the other, they may not be reduced to mean sea level because of the differences between the primary elevations on which the leveling is based. However, prior to and since 1896 precise-level lines have been extended by the Survey and other organizations to more accurately determine inland elevations above sea level, and the elevations originally determined have been corrected from time to time, so that at present nearly all the central points have been reduced to mean sea-level datum and carry with them all the levels resting thereon. As the area referred to a single base is gradually widened, it becomes possible to correct the elevations with reference to a new base, the corrections being, if practicable, entered upon the bench mark.

The location and elevations of the bench marks are published in two ways—first, by a symbol on the atlas sheets, accompanied by the letters "B.M." and the elevation to the nearest foot; second, by lists of bench marks in the annual reports or bulletins of the Survey, with a full description of each bench mark and its exact elevation above sea level to the thousandth of a foot, as adjusted and referred to the various central datum points, these lists being corrected in publications from time to time as better connections are made with sea level. These publications are issued separately for each state.

Geological Survey. As perhaps its basic activity, the Survey is carrying on a survey of the areal geology and geological structure of the United States. By means of this survey it seeks to ascertain the distribution, structural relations, mineralogic character, economic value, and geologic history of the formations of each area surveyed. The results of this survey it publishes in the form of maps, illustrations, and textual descriptions.

The selection of areas for geologic survey is determined by a variety of considerations, corresponding closely to those mentioned in connection with the distribution of the topographic surveys. Still another element is the desirability of giving preference to the public land areas because of the value of geologic mapping in furnishing information needed by the General Land Office in the administration of the land laws.

The degree of refinement of the geologic survey varies with the character of the area and the results desired. The scale of the map on which the results are to be published is determined by the same considerations as in the case of topographic mapping. The three scales used in the topographic mapping—1:250,000, 1:125,000 and 1:62,500—are used also in the geologic mapping and according as one or the other map scale is contemplated the survey is said to be exploratory, reconnaissance, or detailed. Larger scales also are used.

The geologic survey of an area is usually begun after the topographic map has been completed, and that map is used as a base. Occasionally, however, the geologist works with the topographer, or even at times works in advance of formal topographic mapping, but in such cases he is either supplied with some form of map, or himself constructs a rough topographic map as he works.

The field work of the geologist engaged in systematic geologic survey is ordinarily confined to examination of the surface, aided by magnetic observations in connection with magnetic ores. Samples of rocks and ores and specimens of fossils are collected. All mines or other available openings are also examined, but only in exceptional cases are independent borings made for this purpose. The field work is supplemented by office and laboratory work, which includes microscopic and chemical examinations of the rocks and ores collected and paleontologic study of the fossils.

The results of the geologic surveys are given to the public in reports accompanied by geologic maps, diagrams and other illustrations, and in "geologic folios" containing maps and descriptive text. These reports vary widely in scope and character, dealing in some cases with selected areas, in others with selected deposits, and in still others with particular geologic problems. They vary also in size and in the degree of technical knowledge assumed on the part of the reader. The shorter and less technical reports are published under the general name of Bulletins, those of intermediate character under the name of Professional Papers, and the more voluminous and technical treatises under the name of Monographs. The Bulletins and Professional Papers are distributed free for the public, while Monographs are sold.

More recently, the Survey has been issuing, as Bulletins, guidebooks to various parts of the western states, arranged according to the principal tourist routes. Although the geologic features of the country occupy a large part of the text, information is also given, in popular style, relative to the geography, including scenic features, notable historic events, Indian tribes, industrial resources and plants and animals.

A "geologic folio" is thus described by the Survey:[2]

A standard geologic folio comprises a descriptive text, in which the geography, topography, and geology of the area are described; a topographic map; a geologic map, printed in colors, showing the areal distribution of the geologic formations; a geologic map showing the deposits of economic value in stronger or more brilliant colors; a geologic map on which the underground structure of the rocks is exhibited by vertical sections representing what would be seen in deep trenches cut across the area; and a columnar section, in which the rocks are represented in a vertical column in their normal relation one to another, accompanied by a condensed description of their composition, thickness, and relations.

For special areas other maps or illustrations are included. Often it is desirable to present with the description a page of illustrations reproduced from photographs or sketches. For artesian-water areas the distribution of the water-bearing strata, the areas where flowing water may be obtained, and the depths to the water horizon are shown on a special sheet. For active mining regions a map on a scale sufficiently large to show important details of the geology, mines, and ore deposits is inserted. For coal regions of economic importance special methods are used to show the structure or lay of the coal beds—of so much importance to the mine worker. This is done by means of contour lines which show the elevation above sea of the coal bed throughout its occurrence in the area, and then furnish the data for determining the depth of the coal below the surface at any point. Sections of coal beds which have been opened or worked are shown on a large scale and their relation to one another is indicated.

Generally speaking, each folio relates to a single quadrangle, that term having the same meaning as when applied to a topographic map; but some folios cover two or more quadrangles.

Occasionally, the advance of geologic knowledge of a particular area, or the demand for more detailed information, necessitates a revision of maps already published.

The geologic survey now carried on discloses the distribu-

[2] The U. S. Geological Survey: Its Origin, Development, Organization and Operations, p. 99-100, 1904.

tion and thickness of the formations in the area covered, their chemical composition and physical characteristics, and their age—that is, the period of geologic time in which they were deposited, as well as their form—that is, their structure, and the geologic forces which caused their deposition, erosion, and deformation.

The value of facts regarding the mineral character of rock formations for the development of mineral resources need not be dwelt upon. In disposing of the public lands this information is useful not only in shaping policies but in administering the land laws, and such information concerning lands privately owned facilitates and expedites the development of mineral deposits of economic value, or indicates the non-existence of such deposits. In this connection it may also be pointed out that information as to rock formations incidentally includes information as to underground water.

The identification of the geologic positions of several strata in the geologic time-scale is also of value to the economic geologist, as it helps him to determine the probability of the occurrence, in the area, of particular economic deposits. Another value, however, lies in the light it throws on the relation of the geology of the area to that of adjacent areas and, ultimately, of the entire country.

The maps and descriptive texts resulting from geological surveys find a most important use in the hands of the Survey itself in its work of land classification and of geologic research.

It would be difficult to ascertain precisely the total amount spent by the Survey in geologic surveys of the country. The appropriations used have been made under various heads, and some have provided for other work than geological surveys. The total amount appropriated since 1889 for "geologic surveys in various portions of the United States" has been about $5,500,000; that appropriated for paleontologic researches (an item of appropriation no longer in use), and for chemical and physical researches has been over $900,000; and that appropriated for "geologic maps of the United States" has been nearly $2,500,000. To this there should be added a large part of the "lump-sum" appropriations made to the Survey during 1879-1888. It seems fair to conclude, therefore, that the geologic surveys proper, exclusive of surveys in Alaska, have cost nearly $9,000,000.

Investigation of Mineral Deposits. In addition to its surveys of selected areas, the Survey makes many investigations and surveys with a view to determining the occurrence and distribution of the deposits of a particular mineral, or the mineral deposits in a particular area. Much of this work that is done on the public lands is a direct incident to the task of classifying or evaluating those lands under the land laws, and this phase of the work will be discussed in a subsequent section.

The investigations here considered are thus concerned only with a single class of mineral deposits. Though the Survey's work in Alaska has from the beginning been conducted under appropriations for the "investigation of the mineral resources of Alaska," no specific phrase of that kind has ever been used in appropriation bills covering investigations in the United States, which have been conducted under the general authorization for "geologic surveys." Investigations of this type are generally undertaken by the Survey with reference to a deposit of major importance in a field not yet adequately exploited. The Survey attempts to outline the limits of the area in which occurrences of the deposit may be expected, and to indicate the zones in which the richest deposits are likely to be found and the general conditions to be expected. An investigation of this type is the Survey's work on the oil shales of the Green Bay region.

The Survey also, when an unusual demand arises for any mineral product, makes a general investigation of the occurrence of that mineral. Investigations have thus been made of such minerals as monazite, zircon, iridosmine, platinum, and all minerals possessing radio-active and radio-responsive products. More recently a systematic search has been made for new deposits of petroleum, phosphates, nitrates, potash, pyrites, and manganese.

For the proper conduct of exploratory researches of this kind, and indeed for the most effective collection of data relative to mineral deposits in the course of general geologic surveys, it is essential that the geologists conducting the operations be thoroughly versed in the technology of the production of the mineral sought.

Recognizing that the ability to form a correct judgment as to the practical utilization and value of an ore or other mineral deposit in the ground is most effectively acquired through the field observation of operating works and processes for the exploitation and reduction of the same or similar deposits, the Survey regards it of the highest importance that the geologists who may be called upon to appraise undeveloped mineral resources in the public domain or in other regions of the country shall keep as closely as practicable in touch with the mineral industry, and shall have a knowledge of the factors conditioning the successful treatment and utilization of the ores he is examining.[3]

The methods employed in searching for new deposits are not unlike those followed by private prospectors possessed of adequate funds and acting under the advice of trained geologists. If possible, coöperative arrangements are made with private parties who may be drilling in the areas selected for exploration or in adjacent areas by which the cores removed by their drills are turned over to the Survey for examination and report. This method has been employed most consistently in connection with the search for potash deposits which the Survey has been making for several years. Describing this coöperation, the report of the Director for 1915 states:[4]

[3] Report of the Director of the Geological Survey, 1915, p. 47.
[4] P. 44.

Every effort is made to coöperate with all drillers who may be boring in the regions which are under examination or which it is thought may contain large saline deposits. Sample bags and specially prepared log books are furnished to them with the request that cuttings be taken from every "screw," at least in the regions of the formations containing or adjacent to the saline or gypseous deposits. Also samples of brines are solicited. All these samples are tested for potash in the laboratory of the Survey, and the drillers are informed as to whether potash is present in considerable amounts in any of the samples. In the lack of funds sufficient to carry on the exploration more rapidly and especially to conduct boring operations in several areas at once, the effort will be made, whenever possible, to station scientific observers at the sites of new drillings in areas which would appear to be especially promising or in which indications have been reported, in order that the cuttings of the drill may be watched on the ground by a competent observer and tests made of the samples and brines as the work progresses.

In addition to the explorations above described, the Geological Survey is testing all samples sent to it for examination from reported discoveries of potash salts or nitrates. It has also made field investigations of reported deposits concerning which the tests have been favorable, and the information submitted with the samples has warranted the hope that the deposits might prove to be of value.

Research in Geologic and Related Theory. In the Survey's researches for mineral deposits, as in all such researches made by trained geologists for private parties, the area of operations is selected and the operations themselves are guided by a hypothesis or theory relative to the occurrence of the mineral sought. It need hardly be added that the Survey, as the largest single organization engaged in such researches and in the collection of geologic data generally, is itself a prime factor in formulating and developing theories of mineral occurrence.

The Survey has from the first taken the position that a great and peculiar field of usefulness lies before it in this type of research.

The principles controlling mineral deposition and the laws governing the occurrence of ore bodies can be satisfactorily determined only as a result of the comprehensive study of many mines and mining camps. This work is beyond the reach of individuals but is appropriate to the national organization. The application of the principles to particular ore bodies and camps is quite within reach of the individual mining engineers charged with development work.[5]

In large measure geologic and mineralogic data that form a basis for researches of this type already exist in the results of the Survey's work. The search made for specific mineral deposits by the Survey also frequently contributes materials of considerable theoretic value.

The broad and searching observations which should accompany every piece of good economic work comprehend data that are eventually combined in the construction of new scientific hypotheses, some of which, as more observations accumulate, grow into established laws or principles that are in turn of the greatest practical consequence. Thus the detailed studies of the metalliferous deposits in one region or another bring to light evidence from which to determine the genesis of the ores and the modes or conditions of their occurrence.[6]

Both these classes of existing data must, however, frequently be supplemented by special studies in the field and in the Survey's laboratories. As illustrative of the Survey's chemical and physical researches may be mentioned extensive laboratory investigations of the diffusion of gold and silver into lead and tin, the linear force of growing crystals, the migration of natural oil in certain sands, and the enrichment of ores of silver, copper, etc.

The Survey's investigations of certain problems of mineral occurrence border closely upon investigations made by the Bureau of Mines in the field of mineral utilization. The Sur-

[5] The United States Geological Survey: Its Origin, Development, Organization and Operations, 1904. Bulletin No. 227, p. 31 f.

[6] Report of the Director, 1912, p. 16.

vey, for example, has studied the causes of the entrance of water into oil-producing sands, which has occurred on a large scale in certain oil fields, especially in California, with the result that large quantities of oil can never be extracted, and the Bureau of Mines has studied methods of remedying the evil and engineering devices for conserving the oil. In these joint studies the field to be covered by each bureau is clearly outlined and the work is carried out in a spirit of coöperation.

In addition to the investigations designed to develop the theory of the occurrence of some particular mineral, the Survey continuously prosecutes studies and investigations designed to develop general geologic theory. The purpose held in view by the Survey in studies and investigations of this kind was thus clearly stated by Major Powell in his testimony before the joint congressional commission of 1885-86.[7]

> There are many investigations that may not at first seem to lead to economic results, but which ultimately and indirectly are of importance even for economic purposes. The scientific man, especially the geologist, for reasons which I have heretofore set forth, has to do with a vast complexity of problems. To select a part of these problems and work upon them may not lead to substantial results when other problems are neglected that have relation to the first. The whole body of research in geology has a very important practical value, because geologic investigation reveals the wealth buried beneath the surface of the earth. Sound geologic conclusions cannot be reached by following a few narrow lines of investigation, but all such lines of research must be followed that each may shed light upon the other. Unless this principle is fully recognized, a geologic survey might lead to conclusions of no value to the people at large, or conclusions might be reached so erroneous as to be misleading. It therefore often happens that in determining to inaugurate investigations in a particular line the one question asked is, "Are we likely thereby to discover facts that will shed light upon the general

[7] Testimony before Joint Commission, 1885-86, Senate Misc. Decs., 49th Congress, 1st Session, Vol. 4, p. 700.

problems of geology?" feeling assured that ultimately all such research will be of economic value.

In the first decade of the Survey's existence its geologic work was largely devoted to the investigation of the broad problems of American geology, and the formulation of a fundamental geologic classification as the scientific basis for the preparation of the geologic map of the United States. As the major features of this basic problem approached solution, the Survey's attention was increasingly turned to the geological survey of the country. Researches in geologic history and theory have nevertheless been steadily prosecuted. Thus, for some time past the Survey has been engaged in a study of the work of glaciers, the geologic history of the continent during the Quaternary period, and the classification and correlation of the sedimentary rocks of the entire country, work involving the study of all North American fossil faunas and floras and their comparison with those of other parts of the world.

Broad geologic research may involve correlative research in physics and chemistry. Among the researches of this type carried on by the Survey are those relating to questions of terrestrial physics, such as the mechanics of the Panama Canal slides, the variation in the water content of rocks from Panama, the increase of earth temperatures with increase of depth, and the electric activity of ore deposits.

In addition to chemical and physical researches directly related to geologic researches, the Survey from time to time has made researches which may be said to be directed to the elucidation of general physical and chemical problems that have no exclusive relation to the development of geologic theory. Thus, work has been done by the Survey in a revision of the fundamental atomic weights, on the radio-activity of ores, and on the geochemical interpretation of water analyses. Much of the work now done by the Carnegie Geophysical Laboratory was begun in the Geological Survey.

Compilation of Information Relative to Mineral Deposits and the Mineral Industries. In addition to its activities in connection with the exploration of mineral deposits, the Survey collects, compiles and disseminates information relative to all known mineral deposits in the United States, and relative to the production of minerals.

The circumstances under which the Survey began in 1882 its publication of the annual report on the "Mineral Resources of the United States" are set forth in the preceding chapter. In 1905 authority was obtained from Congress (act of March 2, 1905, 28 Stat. L., 960) to publish the separate parts or chapters of this report in pamphlet form as soon as they were transmitted to the printer, it being provided that "the report of the Mineral Resources of the United States shall be * * * printed for each preceding calendar year *as soon as compiled and transmitted for publication,* and that the separate chapters on any given mineral product, such as iron, coal, building stone, and so forth, *shall be printed as rapidly as transmitted for publication.*" This act was passed only to insure prompt publication. "Advance chapters" had been issued long before this act was passed.

In early years, the work of the Survey in this field was confined largely to the compilation of statistics of production. More recently the tendency has been to emphasize the importance of data relative to the occurrence, technology and utilization of the mineral resources of the country. Simultaneously the effort has been to publish more promptly the statistics of production, and at present weekly, monthly and quarterly statistics regarding certain mineral products are being collected.

The collection of statistics showing the production of minerals in the United States involves a complete annual canvass of all the mineral producers, a complete list of whom is kept by the Survey. This canvass is made largely by correspondence, but is supplemented by field work. The technical press is also carefully scrutinized for new names. As a further means of keeping the lists of producers up to date, the lists are printed every year and distributed to persons familiar with the industries for corrections and additions.

In the preparation of its statistical reports for the last six years, the Survey has worked in coöperation with a number of the state geological surveys, the statistics thus collected being used jointly by the state and federal surveys. This coöperative agreement avoids annoyance to the producers by duplicate requests for statistical information and insures more prompt and more complete returns. Eighteen state surveys are now coöperating in this way.

"Mineral Resources of the United States," which for several years has consisted of two volumes, is still the chief medium through which information concerning the country's mineral industries is made public. In addition to this annual compilation and the more frequent shorter periodical statistical statements given largely through the press, the Survey publishes from time to time a report on a particular mineral or group of minerals from material obtained in the course of its investigations. A type of these special compilations is a volume published in 1917, entitled, "Useful Minerals of the United States," giving the location, by states and counties, of the principal deposits of useful minerals, and a glossary showing the composition and character of each mineral, the location of its principal deposits and its chief uses. Among the more technical compilations may be mentioned a 67-page pamphlet on the "Antimony Deposits of Alaska," and a slightly larger paper on "Tungsten Minerals and Deposits." The Survey has also published from time to time in its reports special maps showing the distribution of the deposits of particular minerals.

The Survey's various publications are supplemented by extensive correspondence with the general public. The Survey is at all times ready to give available information to those inquiring into the occurrence and utilization of mineral substances. In this connection the report of the Director for

1916 says that the Survey "is becoming more and more a repository of information concerning the mineral resources of the country, much of which it furnishes through correspondence, giving detailed information which cannot be covered by publications, such as the location of deposits of minerals, both those that are developed and those that are undeveloped, and names of those who can supply the different minerals."

The numerous problems raised since the beginning of the war by the shortage of basic minerals have resulted in an unprecedented demand on the part of other agencies of the government for exact data as to the occurrence, production and utilization of these minerals. This demand the Survey has been able to meet by reason of its unique accumulation, over many years, of data of this type, and its trained organization ready to obtain new classes of data as new demands have been made for them.

Survey of Water Resources. In the course of its topographic and geologic surveys, and of its search for mineral deposits, the Survey, needless to state, collects a vast amount of information relative to the water supplies of the country; and its resulting topographic and geologic maps are of the highest value to one investigating the utilization of the water resources of any area covered by them. The Survey supplements the data collected and published in this connection, which have reference primarily to the location of the surface water supplies, with a survey designed to disclose also their quantity and quality, and to locate sources of underground supply.

The specific investigation of water resources by the Survey dates from the appropriations made in 1888 and 1889 for the investigation of the arid public lands and the extent to which they can be reclaimed by irrigation. The operations under those appropriations and the failure of Congress to make special appropriations for them between 1889 and 1894 were reviewed in the preceding chapter. The investigation of water resources, however, was carried along on a small scale by special arrangements until 1894, when an appropriation authorized its extension to the whole country, and since that time the work has progressed steadily. The maintenance of hydrographic work on its present scale is made possible in large measure by substantial contributions made to the work by many of the states.

The operations embraced in the survey of water resources may be grouped under three heads—the location of supplies, the measurement of their quantity and the investigation of the practicability of their utilization. These operations vary essentially according as they deal with surface or ground waters.

Surface Waters. The location of streams and springs is of course determined by the regular topographic surveys. Their quantity is determined by measuring their flow through a period of years sufficiently long to ascertain definitely the limits of variation, by maintaining stream-gaging stations over all the country, but particularly in the West. During the year ending June 30, 1917, nearly 1,200 such stations were maintained, but only thirty-six of them by the Survey alone, the others being maintained, in coöperation with the Survey, by various bureaus of the government (the Reclamation Service, the Forest Service, the Indian Service, the Weather Bureau, and others) and by states, cities and private persons. At these stations daily measurements of stream height are made, and periodic measurements of the flow or discharge. During the fiscal year 1917 an average of over six measurements of discharge were made per gaging station.

In addition to the data thus collected to determine the quantity of surface waters, analyses are made to determine their suitability for domestic, irrigation, boiler and other uses. In this work samples are usually taken from the stream daily for one year, united in sets of ten consecutive samples and the composite samples analyzed. The results of the analyses, when correlated with records of discharge, show the variations in character of the water with changes in discharge and seasons. They also furnish the basis for classifying the water for irrigation, domestic, and various industrial uses.

For the publication of the results of stream measurements the country has been divided into twelve primary drainage basins, and an annual progress report is published separately for each of these basins. For the publication of the data as to quality of water, utilization of streams, and other special data, no fixed territorial basis is used. Most of the data concerning either surface or ground water are now published in the Survey's series of "Water Supply Papers." Occasionally the study of the water resources of an area for a special purpose, as municipal water supply, results in a publication covering both surface and ground waters in the area studied.

The Survey's investigations of surface waters are put to numerous and important uses. They furnish valuable, and concerning some streams, absolutely indispensable data needed in the consideration and development of irrigation, power, water supply, navigation, and flood protection projects. Their results have been of high value in connection with the irrigation of the arid lands of the West. The location of reclamation projects not only by the national government but by state and private enterprise has for the greatest part been guided by the data gathered by the Survey. Less well known but hardly less important is the part which the Survey's findings have played in aiding power development on the streams of the country.

It is difficult to define in terms of time or of magnitude the entire project of surveying the surface waters of the country, and it is indeed to be questioned whether the work may properly be regarded as definitely terminable, for it is not yet apparent that a point will be reached when the collection of stream measurements and the analyses of water will become unnecessary; on the contrary it is believed that this work, like the collection of climatic data, must be continued indefinitely.

With these limitations in mind, it may be stated that stream measurements covering a period of years are now available for 4,000 points on the streams of the country. The collection of data as to the quality of surface waters has made less progress. Reports including data on the quality of the surface waters in nearly every state have been published, though only six reports cover the quality of surface waters of entire states, and these reports consider only the principal streams. In general, it may be said that most of the principal streams of the country have been investigated for their quality, but these streams constitute a very small percentage of the total surface water supplies of the country.

Ground Waters. The location of ground waters is essentially a geologic problem and is determined principally by geologic methods. It involves an examination of the strata in the region explored and the determination of the position, character, and capacity of those that are water-bearing. These geologic methods give results that are in the nature of forecasts. Drilling is necessary to determine with certainty the yield, head, and quality of the waters sought, but until the fiscal year 1917-18, the Survey did no test drilling for this purpose. The appropriation act for that year provided that of the $175,000 appropriated for water resources investigations, $25,000 might "be used to test the existence of artesian and other underground water supplies suitable for irrigation in the arid and semiarid regions by boring wells." It will be noted that this provision includes only supplies suitable for irrigation. Under this authorization the Survey has begun drilling in a desert area in Nevada where the geologic conditions are believed to be favorable, and has already discovered a large supply of water that will be available for irrigation.

The determination of the quantity of ground water in any given area is difficult. In some areas it is practicable to measure the flow of the ground water by electrical apparatus inserted in test holes or to make estimates based on the observed fluctuations of the water table, which represent the filling or emptying of the underground reservoir. In other areas it is practicable, by observation of surface conditions, to determine approximately the quantity of water that annually enters or is discharged from the underground reservoir. These observations comprise stream measurements to ascertain the quantities of surface waters that percolate into the ground, and measurements of the discharge of ground water through springs and by evaporation from soil and plants. These methods are laborious and difficult and none of them are likely to afford precise results, but they are nevertheless valuable, because they give a tangible basis for estimating the supplies of ground water.

In addition to determining the quantity and quality of the waters a survey of ground waters discloses important factors affecting the practicability of recovering them, such as hydrostatic pressure and the nature of the rock to be encountered in sinking wells. The investigation of these elements is involved in the study of the local geology, which is supplemented by examinations and tests of the existing wells in the area surveyed.

The data gathered as to sources of ground water are naturally of value chiefly in the arid and semiarid regions of the West, where such supplies are needed both for domestic use and for irrigation, but the importance of deep ground water supplies over the whole country is being emphasized. It is highly desirable to obtain supplies of domestic water from sources other than the shallow wells, some of them open, that are sunk near many houses. The water obtained from deep wells has percolated through sands and other material for so great a distance that its impurities have been removed by filtration, and it possesses a sanitary value that cannot well be overestimated, for it is free from the bacteria that cause typhoid fever and the protozoa that cause amœbic dysentery, and its use obviates the necessity for shallow wells that may serve as a breeding place for Anopheles, the mosquito to which malarial infection is due.

Altogether the Survey has accumulated a vast amount of accurate data of both scientific and economic value on the ground waters of the United States. A total of 474 papers has been published relating in some way to ground water, of which 271 papers relate primarily to this subject. Many of these papers include detailed maps showing the ground-water conditions in specific areas. In addition to the data already published there is a great quantity on file at the Geological Survey still unpublished.

About 75 per cent of the area of the United States (excluding Alaska and the Insular Territories) has been covered by ground-water surveys of some sort, and about 25 per cent has been covered by what may be called standard ground-water surveys. The total area covered by intensive surveys is very small, amounting to only a fraction of 1 per cent of the total area of the United States.

The Survey for some years has published in its water supply papers notes on desert watering places, but in 1917 a specific appropriation of $10,000 was granted by Congress "for discovering, developing, protecting, and rendering more accessible, springs, streams, and water holes on arid public lands of the United States; for erecting and maintaining suitable and durable monuments and sign-boards; and for providing convenient and ready means, apparatus, and appliances by which water may be brought to the earth's surface." The Survey has devoted the funds thus appropriated to prosecuting more vigorously than has hitherto been possible the survey of watering places, and to preparing guides and maps, and erecting sign-posts directing to them.

The total area sufficiently arid to require maps and guides to desert watering places is about 570,000 square miles, or nearly one-fifth of the area of the United States, exclusive of Alaska and the Insular Territories. Of this area about 75,000 square miles have been surveyed since July 1, 1917, and additional areas of approximately the same aggregate extent are covered by older guides to desert watering places. The region to be covered lies east of the Cascade and Sierra Nevada Mountains and west of the following cities: Boise, Twin Falls, Salt Lake City, Grand Junction, Santa Fé, Roswell, Pecos, and Fort Stockton. The area has been tentatively divided into twenty-six districts, for each of which a guide is ultimately to be prepared. The preparation of maps and guides and the erection of a fairly adequate number of sign-posts for the entire area will cost about $100,000. It is hoped that this work can be done within the next few years.

Investigation in Problems of Water Utilization and Control. From time to time the Survey undertakes the investigation of large problems of water utilization and control. The early investigations made by the Survey into the problem of irrigating the arid lands of the West, and the more systematic work of the Powell Irrigation Survey, in 1888-90, have already been mentioned. It is worthy of note that in his report for 1888, in discussing the problem of irrigation, Major Powell adverted also to the possible reclamation of the flood plains of the lower Mississippi valley by the control of the headwaters of that stream and its chief tributaries. A problem closely akin to this was brought expressly within the province of the Survey by the act of 1908, to which more extended reference will be made in the next section, imposing upon the Survey the duty of examining forest lands proposed to be purchased by the government for the prevention of floods on navigable streams.

While one or another problem of water utilization or control has thus from time to time been brought within the purview of the Survey's activities, no express authorization to investigate generally problems of this type has ever been conferred by statute. Such investigations as are currently carried on are conducted under the general authorization to investigate "the water resources of the United States."

Illustrative of the type of investigation undertaken by the Survey in this direction are the continuing investigations of the problems of flood prevention and of reclamation of marsh and swamp lands, and the special investigation of the hydraulic mining débris problem. The investigation last mentioned, one of the most extensive and original of those made in this field by the Survey, was undertaken at the request of the California Miners' Association with a view to devising measures whereby the interests of agriculture and navigation in the streams carrying mining débris could be harmonized with those of the placer mining industry. The investigation by the Survey, which extended over more than ten years, involved research in the existing literature on the subject, laboratory study of the laws of transportation of detritus by running water, and a field examination of the entire course of the Sierra Nevada River. The report of this investigation, which was published in 1917, recommends a coöperative control by all the major users of the Sierra's water whether for mining, irrigation, or power development.

Examination Under Act of March 1, 1911, of Land Proposed to be Purchased by the Government for Protection of Navigable Streams. The act approved March 1, 1911, known as the Weeks Act, which provides for the acquisition of land for the purpose of conserving the navigability of navigable streams, imposed a new duty on the Geological Survey. The lands to be purchased must be selected by the National Forest Reservation Commission, but under section 6 of the act they may be purchased only after the Geological Survey has examined them and submitted a report on them to the Secretary of Agriculture "showing that the control of such lands will promote or protect the navigation of streams on whose watersheds they lie." The law thus places upon the Geological Survey the responsibility of determining the fundamental question whether the control of any specific tract of land will promote or protect navigability, and stipulates that a favorable report to the Secretary of Agriculture must precede the purchase of the tract considered.

Following is a description of the methods employed in beginning this work as contained in the annual report of the Director of the Survey for 1911:

On the initiative of the Survey a conference was arranged between representatives of the Department of Agriculture and of the Department of the Interior, and an agreement was reached concerning procedure in the administration of this new law, so far as the Forest Service and the Geological Survey are concerned, to the end that the examinations of land by the two bureaus might be coördinated. In this agreement, which was approved by the two Secretaries on May 3, 1911, it is set forth that the examination by the Geological Survey will include the determination of the relation of the headwater streams to the navigable streams to which they are tributary, the local observation of the headwater stream or streams draining the tract or tracts in question with reference to run-off characteristics and to nature and amount of suspended material, the classification of the surface formations of the tract with reference to permeability and storage capacity and to resistance to erosion, and the securing of such additional topographic data, in coöperation with the Forest Service, as are needed by the two bureaus in their examination of the tract.

The greater part of the examinations necessitated by this act were completed within the first two years after the passage of the act, but other examinations have since been made from time to time.

Investigation of the Character and Value of Public Lands Made Necessary by the Public Land Laws. As has been seen in the preceding sections, the surveys and investigations made by the Survey, though they extend over the whole country, have in large measure related to what may be termed the public land region of the United States. Indirectly, therefore, the Survey's general work has had and continues to have a far-reaching influence upon the development of the public lands. In addition, however, the Survey aids materially in the actual administration of the public land laws, for it is called upon to investigate the mineral and water resources of specific areas of the public land, as well as to determine the value of certain tracts, and to recommend to the Secretary of the Interior or the General Land Office the designation or non-designation of each tract in such areas as falling under one or another of the statutes governing the public lands. Moreover, on occasion, it is charged with the drafting, for the approval of the Secretary of the Interior or the President, of further regulations for the administration of these laws.

Origin and Development. For a decade or more before the creation of the Survey, Congress, by legislation for the disposition of the public lands, had in increasing degree drawn distinctions between the several types of land. The homestead law of 1862 had expressly excluded mineral lands from its provisions. Every grant to a railroad, beginning in 1862 and ending in 1871, excluded from the grant all lands containing minerals other than coal and iron. In the decade 1865-1875 were enacted the basic provisions governing the disposition of mineral lands. In 1873 was enacted the coal land law, which provided not merely for the sale of coal lands under special provisions, but, inferentially at least, for their evaluation before sale. In 1877 still another novel distinction was introduced into the land laws by the enactment of a provision for the sale of desert lands.

Despite the existence of these special provisions relative to the disposition of desert, timber, mineral and coal and iron lands, neither the statutes nor the machinery of the Land Office made any provision for the independent determination by the government, in advance of sale, of the character of the lands to be sold. The sole method employed was that of proof offered by the entryman or claimant and subject to challenge before the officers of the Land Office.[8] The results

[8] "At present the General Land Office possesses the machinery for the survey, classification, and sale of the public lands. In that bureau the field-notes and maps of the various deputy surveyors are intended

secured by this method were thus characterized by Major Powell, subsequently Director of the Geological Survey, in 1878, the year before the Survey was created: "No adequate provision is made for securing an accurate classification, and to a large extent the laws are inoperative or practically void; for example, coal lands should be sold at $10 or $20 per acre, but, the department having no means of determining what lands belong to this class, titles to coal lands are usually obtained under the provisions of statutes that relate to lands of other classes—that is, by purchasing at $1.25 per acre, or by homestead or preëmption entry."

Major Powell was accordingly of the opinion that "an examination of the laws will exhibit this fact—that for the classification contemplated therein a thorough survey is necessary, embracing the geological and physical characteristics of the entire public domain." The National Academy of Sciences, to which Major Powell had expressed this opinion, and which, as already stated in the outline of the origin of the Survey, had been requested by Congress to investigate the entire subject of government surveys, reported that "to meet the requirements of existing laws in the disposition of the agricultural, mineral, pastoral, timber, desert, and swamp lands, a thorough investigation and classification of the acreage of the public domain is imperatively demanded. The Land Office shall also call upon the United States Geological Survey for all information as to the value and classification of lands."

In the act creating the Geological Survey, as already stated, Congress provided that it should have charge of the "classifi-

to convey sufficiently accurate information for the general guidance of the officers who execute the sales. The law also provides a method of proof as to the character of lands, which forms an indispensable stage in the process of sale. Any transaction as to a piece of public land may be challenged before the proper officers, and its character may be determined by competent proof. The present method of sale of the public lands depends, therefore, chiefly upon a rule of law rather than the classification of experts in advance of the procedure of sale." First Annual Report of the Geological Survey, 1880, p. 5.

cation of the public lands." In the same act, however, provision was made for the creation of a commission of five members, of whom the Director of the Geological Survey was to be one, to report to Congress "first, a codification of the present laws relating to the survey and disposition of the public domain; second, a system and standard of classification of public lands, as arable, irrigable, timber, pasturage, swamp, coal, mineral lands, and such other classes as may be deemed proper, having due regard to humidity of climate, supply of water for irrigation, and other physical characteristics; third, a system of land-parceling surveys adapted to the economic uses of the several classes of lands; and, fourth, such recommendations as they may deem wise in relation to the best methods of disposing of the public lands of the western portion of the United States to actual settlers." The commission appears to have given consideration to the practicability of a classification of the public domain in advance of disposition, and to have decided adversely to any change in the existing practice in this regard.

Accordingly, the Director of the Survey stated in his first annual report:

Upon examination of the existing land system, I have assumed that Congress, in directing me to make a classification of the public lands, could not have intended to supersede the machinery of the Land Office, and substitute a classification to be executed by another bureau of government, without having distinctly provided for the necessary changes within the Land Office, and adjustment of relations between the two bureaus.

I have therefore concluded that the intention of Congress was to begin a rigid scientific classification of the lands of the national domain, not for purposes of aiding the machinery of the General Land Office, by furnishing a basis of sale, but for the general information of the people of the country, and to produce a series of land maps which should show all those features upon which intelligent agriculturists, miners, engineers, and timbermen might hereafter base their operations, and which would obviously be of the highest value for all students of the political economy and resources of the United States. Studies of this sort, entirely aside from the administration of the Land Office, can be made of the highest practical value.

The position thus taken at the beginning of the Survey's history was acquiesced in by Congress, which took no steps to modify the wording of the statute. Special duties, involving the administration of the public lands, were indeed from time to time temporarily imposed upon the Survey. Such were the duties exercised in 1888-91 with respect to the reservations of reservoir sites and irrigable lands, in 1896-1905 in the examination of the forest reserves, and in 1902-07 in the administration of the reclamation act. Aside from these exceptional functions, however, it may be said that until about 1906 the Survey remained almost wholly detached from the administration of the land laws.

In the years 1905 and 1906 the general public began to realize that large areas of valuable coal lands in the West had been obtained from the government by means of agricultural entries. The frauds thus perpetrated were so great as to shock the public mind and to call for some immediate action to prevent further similar looting. Accordingly, the President, on June 29, 1906, directed the Secretary of the Interior to withdraw from entry all valuable coal lands. The Survey had previously been making special studies of certain western coal fields and its geologists had assisted in unearthing some of the coal frauds. It was, therefore, prepared to submit a list of lands which should be withdrawn from entry, and on July 26, 1906, the Acting Secretary of the Interior withdrew from all forms of entry all the lands on the list submitted by the Survey. There was thus set in motion a train of legislative and administrative action extending over several years by which the entire system of disposing of the public coal lands was revised and placed upon the basis of a valuation by the Survey before disposition. The work of evaluating these lands is still in progress by the Survey.

In the following year the Survey made recommendations to the Secretary for the withdrawal of oil and gas lands, and in 1909 of potash lands pending legislation for their disposition. In the same year the Survey's activities in this field were extended to lands containing water-power sites and in 1912 to public water reserves, that is, water supplies necessary for the utilization of the public range, and still later to phosphate lands, oil-shale lands, and lands suitable for designation under the enlarged homestead laws. All this classification work was undertaken, and is still continued, without express legislative warrant other than that contained in the original act of 1879. In 1917, however, Congress made an appropriation of $150,000 "for the examination and classification of lands requisite to the determination of their suitability for enlarged homesteads, stock-raising homesteads, public watering places, and stock driveways, as required by the public land laws,"—the first specific appropriation for land classification work ever made to the Survey.

Legal Provisions Applied in Land Classification. For an understanding of the methods employed by the Survey in land classification work, some knowledge is necessary of the provisions of law towards the execution of which that work is directed. These provisions are too numerous and intricate for analysis here, but in their relation to the classification problem of the Survey they may be said to fall into two classes—those involving a determination of the mineral resources of the particular tract involved, and those involving a determination of its relation to the available water supply. In the first class fall those provisions requiring a determination as to whether the land contains minerals. If it does it is not open to homestead entry nor may it be taken under a railroad grant (unless it contains only coal and iron) or under a grant to a state. It is open to entry only under special laws, which, however, vary according as the land contains coal or other minerals. In general, these special provisions affect only the extent and character of the right acquired by the entryman or patentee and the requirements of settlement or payment to be met by him. But in the case of coal lands a special provision permits and inferentially requires the valuation of the lands previous to their sale.[9]

The provisions of law involving a determination of the relation of the tract in question to the available water supplies are of three distinct classes—those relating to entries upon arid tracts by individuals, those relating to rights of way for irrigation, power or water supply purposes, and those relating to grants of irrigable lands to the states.

Under the desert land laws entries are permitted upon lands not capable of producing crops without irrigation, while under the enlarged homestead laws homestead entries of 320 acres may be designated by the Secretary of the Interior upon lands not susceptible of irrigation at a reasonable cost from any known source of water supply; and additional privileges are extended in certain cases to tracts which have not upon them such a sufficient supply of water suitable for domestic purposes as would make continuous residence upon them possible. The act of December 29, 1916 (39 Stat. L., 862) authorizes the Secretary of the Interior to designate as "stock-raising lands," open to homestead entries of 640 acres, lands chiefly valuable for grazing and for raising forage crops.

Various acts provide for the grant of right-of-way permits over the public lands for canals, ditches and reservoirs for irrigation and for mining and milling purposes; for power development and transmission, and for municipal water sup-

[9] The law (sec. 2347 of the Revised Statutes) provides that "every person above the age of twenty-one years, who is a citizen of the United States, or who has declared his intention to become such, or any association of persons severally qualified as above, shall, upon application to the register of the proper land office, have the right to enter, by legal subdivisions, any quantity of vacant coal lands of the United States not otherwise appropriated or reserved by competent authority not exceeding one hundred and sixty acres to such individual person, or three hundred and twenty acres to such association upon payment to the receiver of not less than ten dollars per acre for such lands where the same shall be situated more than fifteen miles from any completed railroad, and not less than twenty dollars per acre for such lands as shall be within fifteen miles of such road."

plies. Action upon application for rights of way for these purposes obviously requires a determination of the feasibility of the project and the limits of the right of way required.

Under the Cary Act the Secretary of the Interior is authorized to contract with any state in which there are arid public lands for the reclamation of such lands by the state, and for their patent to the state upon satisfactory proof that an adequate supply of water for reclamation has been developed.

In the administration of the foregoing provisions it is frequently necessary also to apply the provisions of the so-called separation acts, the purpose of which is to permit the alienation separately of the values of a tract for the several purposes—agricultural, mineral or water—of which it may be susceptible of use.[10]

As against all these acts making provision for the disposition of the public lands, there is the general act of June 25, 1910 (36 Stat. L., 847) providing for the withdrawal from entry of any of the public lands in the discretion of the President. It is provided that "the President may at any time, in his discretion, temporarily withdraw from settlement, location, sale, or entry any of the public lands of the United States, including the District of Alaska, and reserve the same for water-power sites, irrigation, classification of lands, or other public purposes to be specified in the orders of withdrawals." By another act of the same date similar, though more limited, provision was made for withdrawal of lands within Indian reservations. Previous to the enactment of these acts, however, the power of withdrawal of lands from entry had long been exercised by the President without specific statutory warrant.

[10] In addition there might be mentioned the reclamation act, the application of which requires a determination in connection with each project, not merely of the lands irrigable by the proposed works, but of the size of tract within each portion of such lands required for the support of a family. Special acts also grant rights of way for irrigation,, power and municipal water supply purposes over the national forests. These acts are administered, however, by the Reclamation Service and Forest Service, respectively, without reference to the Geological Survey.

The principal purpose for which the power of withdrawal is now exercised, besides that of classification, is that of reserving from alienation public lands of special value, for the disposition of which Congress has made no adequate provision. At the present time oil, gas, potash and phosphate lands, and water-power and reservoir sites are being withheld from entry pending legislation. Some of the tracts involved have been thus withheld from entry since 1908.

The power of withdrawal may also be exercised to create permanent reserves in the public lands. Thus, several so-called public water reserves have been created in the great semiarid grazing areas of the West for the protection of the public range. In addition large petroleum reserves have been set up to insure an adequate supply of fuel oil for the Navy.

The determination of withdrawals, whether permanent or merely pending legislation, obviously requires an examination and classification of the lands to be reserved in the same manner as if the lands were to be thrown open to entry.

Under present practice, the work of investigation and evaluation carried on by the Survey is of two distinct kinds—the one the systematic and comprehensive examination and classification of large areas, the other merely that of rendering to the Land Office reports upon particular cases before it.

Systematic Classification. In the systematic classification of an area there is generally sought to be determined not the classification of each tract under each of the land laws but only under some particular provision of law, or for some specific contemplated withdrawal. Thus, the examination of an area may be undertaken solely to determine which of the tracts involved are phosphate lands. It may thus be found necessary, from time to time, to reëxamine and classify, with respect to a particular provision, an area already examined one or more times for classification with respect to other provisions.

Up to the present time, systematic classifications have been attempted of coal, oil and gas, phosphate and potash lands, of lands suitable for enlarged and stock-raising homesteads, of water-power and reservoir sites, and of public water reserves. A systematic classification of oil-shale lands has been undertaken and is still in progress, such lands being, however, classified only as mineral. No systematic classification has been attempted of mineral lands generally [11] or of desert lands.

The first step in the systematic examination and classification of an area thought likely to contain mineral or water resources, bringing it under special provisions of law, involves several distinct stages—that of withdrawal of the lands from entry pending classification, that of examination and classification, and, in the case of coal lands, evaluation, and that of restoration of the lands to entry.

Withdrawals for purposes of classification are recommended by the Survey as soon as it finds probable cause for believing that a given area contains lands falling under any of the special designations above referred to as calling for systematic classification. The facts upon which such belief is based may result from the Survey's own explorations or may be communicated by private parties, or may come to the attention of the Survey through news of the discovery of mineral deposits in areas not hitherto thought to contain such deposits.

The recommendations of the Survey as to withdrawals are made direct to the Secretary of the Interior, who, in turn, refers them to the President with his recommendations.

The extent of field examination found necessary for the classification of an area depends of course upon the amount of data already available to the Survey as a result of investigations previously made by it in the area, whether for classification or for other purposes, or those which may have been made by the field force of the General Land Office. Generally speaking, a previous systematic geological survey of an area

[11] By act of June 25, 1910 (36 Stat. L., 739) special provision is made for the classification of the lands within the Northern Pacific Railroad grant with respect to their mineral or non-mineral character. Work under this provision has been in progress since 1910.

will not obviate the necessity of field examination for classification.

As soon as possible after a withdrawal has been made, detailed examination of the withdrawn area is begun.

The classification of the individual tracts is controlled in large measure by standard regulations previously formulated by the Survey.

The regulations applying to the several types of mineral land generally define the amount of the given mineral which the tract must be estimated to contain to warrant its classification as such land. The regulations applying to coal lands necessarily go further and prescribe in the greatest detail the factors to be considered in evaluating such lands.

The technical methods used in the field examinations made for classification necessarily vary in accordance with the mineral or water supply whose presence is in question; and as they have been very fully set forth by the Survey in a publication issued in 1913,[12] they will not be described here.

In general, the results of the Survey's classifications are recorded in such form as to be available only for the use of the Survey, the General Land Office and the Office of the Secretary of the Interior. For its enlarged homestead designations, however, the Survey has printed maps showing the tracts designated.

As soon as an area has been classified, recommendation is made to the Secretary for the restoration to entry of the lands found not to fall within the provisions of the law. The lands falling within such provisions are either restored to entry under those provisions or are reserved from entry under the withdrawal act.

The classifications made by the Survey are accepted by the General Land Office as controlling, with respect to certain types of classification by order of the Secretary and with respect to others by practice. The classification by the Survey

[12] U. S. Geological Survey, Bulletin 537, The Classification of the Public Lands, p. 50 ff.

does not, however, in any way impair the legal right of an entryman to contest before the Commissioner of the General Land Office or the Secretary the propriety or accuracy of the classification.

The progress up to June 30, 1917, of the systematic classification of lands classified and restored to entry under such classification may be seen from the following:

Lands classified as	Acres
Coal lands	21,226,708
Oil-shale lands (classified as mineral lands)	3,992,036
Mineral lands in Northern Pacific Railroad grant in Idaho and Montana	224,218
Lands suitable for enlarged homesteads	275,633,861

The status on June 30, 1917, of lands withdrawn from entry either permanently or pending the enactment of legislation for their disposition is as follows:

Withdrawn as	Acres
Oil lands	6,348,640
Potash lands	130,629
Phosphate lands	2,506,398
Water-power sites	2,429,097
Reservoir sites (valuable principally for the storage of water for irrigation)	80,251
Public water reserves	201,708

These figures, it should be noted, include under each class all lands now withdrawn from entry. The lands so withdrawn, however, have not all been examined and classified, and are, therefore, subject to restoration to entry should they be found, on examination, not to fall in the class indicated.

Classification on Applications. The classifications made by the Survey of tracts with respect to which entries or applications have been filed in the Land Office now cover virtually all applications for and entries upon lands not yet classified, except homestead and desert land entries, metalliferous mineral entries, and coal land applications, the investigation of which, both in office and field, is made solely by the General Land Office.

In its land classification work on applications the Survey makes no field examinations but merely renders an opinion to the General Land Office on the basis of the data already on hand. If these data be inadequate, the necessary field examination is made by the General Land Office itself.

Operation of Lithographing and Engraving Plant. The operation of a lithographing and engraving plant is listed as one of the activities of the Survey, since this plant is operated not only for the execution of work for the Survey itself, but for other branches of the government as well. The policy of executing work for other services of the government has been adopted in order that the plant of the Survey may be maintained on the scale necessary to perform economically its own special work, to keep its equipment and personnel more continuously employed, and to make possible the concentration of a large force for emergency work for the Survey and for other government bureaus or offices as the demand may arise. The report of the Director of the Survey for the year 1917 shows that a large quantity of lithographic work of great variety was performed for the Government Printing Office in printing maps to be incorporated in publications of many departments of the government. Much work was also done directly for other departments and bureaus. The value of the work thus done directly or indirectly for other branches of the government amounted to about $97,000, for which the Survey's appropriation for engraving and printing geologic maps was reimbursed by transfer of credit on the books of the United States Treasury.

A small amount of work, totaling about $550, involving the reproduction, for publication, of Survey maps, was also done for eleven organizations and companies. The money received for this work is turned over to the Treasury to be credited to miscellaneous receipts.

The total number of copies of all kinds of engraved and printed matter produced during the year 1916-17 was 3,338,028, requiring 9,805,217 impressions.

CHAPTER III

ORGANIZATION

The primary divisions of the Survey are six, and are known as branches. They are:

1. Administrative Branch
2. Topographic Branch
3. Geologic Branch
4. Water Resources Branch
5. Land Classification Board
6. Publication Branch

Administrative Branch. The Administrative Branch includes all divisions carrying on all the general business of the Survey that is not included in the technical, scientific and publication work. It is composed of two offices:

1. Office Proper of the Director
2. Office of Chief Clerk

The Director is by law the administrative and scientific head of the Survey, reporting officially to the Secretary of the Interior, directing all scientific work, and being generally responsible for the organization and operation of the service. The Director spends a part of his time in the field exercising a general oversight of the Survey's field work. In his absence his administrative duties are performed by a geologist who is designated as Acting Director.

The chief clerk is the officer having direct charge of, and responsibility for, the conduct of the administrative work of the service. For this work his office embraces five divisions which, with their sections, are as follows:

1. Office Proper of Chief Clerk
2. Executive Division
   a. Office of Chief of Executive Division
   b. Mails, Records and Files
   c. Appointments
   d. Addressograph, Multigraph and Mimeograph Section
3. Division of Accounts
4. Library
5. Messenger Force

The Executive Division has charge of appointments, mails, records and files. The Division of Accounts has supervision of administrative bookkeeping, property accountability, purchases, express and freight. Of the 934 permanent employees of the Survey on June 30, 1917, 60 were employed in the Administrative Branch.

Topographic Branch. The Topographic Branch conducts the topographic surveys and prepares the originals of the topographic maps. The manuscripts of the several publications on spirit leveling and other topographic subjects are also prepared by this branch. The branch is administered by the chief geographer.

On June 30, 1917, the Topographic Branch comprised 166 of the 934 employees of the Survey.

For purposes of supervision and administrative control, the branch is divided into seven divisions:

1. Administrative and General
2. Atlantic Division
3. Central Division
4. Rocky Mountain Division
5. Northwestern Division
6. Pacific Division
7. Division of Military Surveys

Each of the five territorial divisions has a geographer in charge who reports to the chief geographer. Each geographer in charge has supervision of all field and office work on the topographic maps prepared in his division. Each division maintains headquarters in the field during the field season, the chief geographer and most of his force returning to the Washington office at its close for the execution of work of preparing the maps for publication. Inspectors, whose functions are to examine field work in progress and to aid in the maintenance of high and uniform standards, assist the chief geographer in the technical supervision of the work of the branch.

The operations of this branch do not extend to Alaska, all topographic work in that territory being carried on by the Division of Alaskan Mineral Resources in the Geologic Branch.

Geologic Branch. The Geologic Branch of the Survey is under the charge of the chief geologist. The number of employees under his direction on June 30, 1917, was 245, the largest number employed in any of the six branches of the Survey. In addition to this number 61 persons, most of them geologists, were employed occasionally by this branch. It is organized in four divisions:

1. Division of Geology
2. Division of Chemical and Physical Research
3. Division of Mineral Resources
4. Division of Alaskan Mineral Resources

Division of Geology. The Division of Geology is under the immediate direction of the chief geologist, who is also head of the Geologic Branch.

The division is organized into sections corresponding to the several types of field and research work it performs. These sections do work in areal, glacial and coastal plain geology, paleontology, and metalliferous, non-metalliferous, coal, oil, and gas deposits. Each section maintains its own field parties, which, while giving attention in their examination primarily to the special province of the section, also gather data relative to the general geology of the areas examined. Thus, while the sections on areal geology are primarily responsible for the determination and mapping of the areal geology of the regions they examine, they receive assistance from other sections in their work in areas with which those sections come into special contact. The sections of the division are as follows:

1. Administrative and General
2. Section of Eastern Areal Geology
3. Section of Western Areal Geology
4. Section of Coastal Plain Investigations
5. Section of Glacial Geology
6. Section of Paleontology and Stratigraphy
7. Section of Geology of Metalliferous Deposits
8. Section of Geology of Non-Metalliferous Deposits
9. Section of the Geology of Eastern Coal Fields
10. Section of the Geology of Western Coal Fields
11. Section of the Geology of Oil and Gas Fields

The two sections of Areal Geology perform reconnaissance and detailed work in areal or general geology, in pursuance of the plan to complete the geologic atlas of the United States. The Sections of Eastern and Western Areal Geology make investigations east and west of the one hundredth meridian, respectively.

The Section of Coastal Plain Investigations carries on a study of the geology of the Atlantic and Gulf coastal plains of the United States, with special reference to their geologic history, giving particular attention to investigations of the stratigraphy, paleontology, and origin of the various formations.

The Section of Glacial Geology is engaged in the study of the glaciated area, the mode of action and the results of the action of glaciers, the glacial and interglacial deposits and the contemporary deposits of the regions bordering the glaciated area, as well as the geologic history of the continent during the Quaternary period.

The Section of Paleontology and Stratigraphy investigates and determines the relative age and equivalence of the strata in different areas, and refers the formations to a geologic time scale.

The Section of Geology of Metalliferous Deposits studies metalliferous deposits and minerals and investigates the conditions and modes of ore deposits; while the Section of Geology of Non-Metalliferous Deposits studies non-metallic deposits and minerals, exclusive of fuels. The Sections of Geology of Eastern Coal Fields and Geology of Western Coal Fields investigate coal deposits east and west of the one-hundredth meridian, respectively, and the Section of the Geology of Oil and Gas Fields makes examinations of oil and gas fields.

The Committee on Geologic Names is a standing committee organized to insure uniformity in geologic classification and nomenclature in Survey publications, to prevent unnecessary duplication of geologic names, and to reduce if possible the number of names employed. Under the instructions of the Director this committee considers all geologic names used in manuscripts prepared by members of the Geological Survey for both official and unofficial publication. The chiefs of the sections in the Division of Geology and the chief of the Division of Alaskan Mineral Resources serve as associate members of this committee and attend meetings at which matters pertaining to their respective sections or divisions are under consideration.

The Committee on Physiography, as stated in the report of the Director for the year 1916, is charged with the following duties:

1. To read critically and give advice as to physiographic papers or physiographic chapters or sections in other papers submitted for publication by the Survey.

2. To consider the classification and nomenclature of physiographic provinces.

3. To prepare or make recommendations as to the preparation of physiographic descriptions in popular language, to be printed on the backs of topographic sheets.

4. To consult with geologists regarding the solution of physiographic problems.

5. To formulate the usage of physiographic terms.

6. To confer with the chiefs of sections and with the chief geologist regarding physiographic work to be undertaken by the Geological Survey.

Division of Chemical and Physical Research. The Division of Chemical and Physical Research, under the supervision of a geologist in charge, makes such chemical and mineralogical determinations as are required by the general work of the Survey and carries on special research work in chemistry and physics. It is organized in three sections:

1. Administrative and General
2. Section of Chemistry
3. Section of Physics

Division of Mineral Resources. The Division of Mineral Resources, under the direction of a geologist in charge, makes field and office studies of mineral resources and industries and collects and compiles for publication statistical information concerning the mineral industries of the United States. It is organized in the following subdivisions:

1. Administrative and General
2. Metals Section
3. Non-Metals Section
4. Field Offices

The Metals Section and the field offices are in charge of the head of the division.

Field offices are maintained at Denver, Salt Lake City, and San Francisco. To each of these offices is assigned the collection and compilation of the metal-mine reports of a given part of the western territory. The heads of these offices act in coöperation with other members of the Survey when necessary and by keeping in close touch with the mining regions of the West they procure and disseminate much valuable information that could not readily be obtained otherwise. These offices are, therefore, general Survey headquarters and bureaus of information in their respective regions.

The reports issued by the division on the occurrence and production of the several minerals are prepared in part by members giving all their time to the division, and in part by other members of the Survey assigned for part of their time to the work. The supervision of the work remains throughout in the hands of the division.

Division of Alaskan Mineral Resources. The Division of Alaskan Mineral Resources, under the direction of a geologist in charge, investigates the mineral resources of Alaska. The work of this division includes the making of both topographic and geologic surveys and the investigation of water resources, as well as the collection and compilation of the statistics of the mineral production of Alaska.

During the fiscal year 1916-17 twelve parties were engaged in surveys and investigations in Alaska. Seven of these parties were engaged in geologic surveys, three in topographic surveys, one in topographic and geologic work combined, and one in stream gaging.

Water Resources Branch. The Water Resources Branch is administered by a chief hydraulic engineer, who reports to the Director. This branch conducts surveys of the water resources of the United States, makes field examinations of the water supplies of the public lands in relation to the several public land laws, investigates problems of water utilization, and studies the ground water supplies of the arid public lands. In this branch are prepared for publication all the water supply papers published by the Survey.

Of the 934 permanent employees of the Survey on June 30, 1917, 102 constituted the personnel of the Water Resources Branch. The branch is organized in six divisions:

1. Administrative and General
2. Division of Surface Waters
3. Division of Ground Waters
4. Division of Water Utilization
5. Division of Enlarged and Stock-Raising Homesteads
6. Division of Quality of Water.

Division of Surface Waters. The Division of Surface Waters, under the direct supervision of a hydraulic engineer, is engaged primarily in measuring the flow of rivers. It also carries on special investigations of conditions affecting stream flow and the utilization of streams.

At selected places, known as gaging stations, measurements of discharge are made and other data are collected from which the daily flow of streams is computed. On June 30, 1917, there were 1,178 gaging stations, only 36 of which, however, were maintained by the Survey alone, all the rest being maintained in coöperation with other services of the government, with private persons, and with states and municipalities.

The work of the gaging stations, as well as all other field investigations incident to the work of the division, is supervised from district offices, of which there are 17, including one in Hawaii. Data collected from the district offices are transmitted to the Washington office, where they are reviewed in the computing section and prepared for publication. This review insures accuracy in the data and brings the results from different parts of the country to a uniform standard. Regular field inspections are made, and through annual conferences of the district engineers the work in the several districts is further standardized.

Division of Ground Waters. The Division of Ground Waters, under the administration of a geologist, makes surveys of waters occurring below the surface of the ground with reference to their utilization. It undertakes the study of conditions controlling the quality and quantity of underground waters, and of methods of recovering and utilizing such waters. Most of the members of the division are geologists and close coöperation exists with the Geologic Branch.

Division of Water Utilization. The Division of Water Utilization is under the direct administrative control of the chief of the branch. It makes all field examinations necessary for the classification of the public lands with respect to their water resources, except as to their suitability for designation as enlarged or stock-raising homesteads. It makes all field examinations incident to the report to the Secretary by the Survey on applications for rights of way over the public lands, for power development, irrigation, or municipal water supply.

Division of Enlarged and Stock-Raising Homesteads. The Division of Enlarged and Stock-Raising Homesteads is under the direct supervision of the chief of the branch. Its work embraces the examination of public lands thought suitable for designation as open to entry under the enlarged homestead or stock-raising homestead laws.

Division of Quality of Water. The Division of Quality of Water is under the direct administration of a chemist. It investigates the mineral quality of surface and ground waters and analyzes the samples of water collected by the field forces of other divisions.

Land Cassification Board. The function of the Land Classification Board is to classify and evaluate the public lands under the public land laws, and to give opinion on applications for entry or right of way on such lands.

In 1908, in consequence of the growth of the Survey's land classification work in volume and importance, the Land Classification Board was organized as a section of the Geologic Branch. Upon this board was laid the duty of systematizing that work more completely and of adopting certain definite principles that might be consistently, uniformly and thoroughly applied to it. On May 1, 1912, the importance of the classification work was further recognized by the designation of the Land Classification Board as a separate branch of the Survey, coördinate in rank with the Geologic, Topographic and Water Resources Branches.

The board maintains no field staff of its own, but bases its reports on the results of field and laboratory investigations made by other branches of the Survey or by the General Land Office.

The board is in charge of a geologist. It is composed of three divisions, as follows:

1. Administrative and General
2. Division of Mineral Classification
3. Division of Hydrographic Classification

Of the 934 permanent employees of the Survey on June 30, 1917, 42 constituted the personnel of the Land Classification Board.

Division of Mineral Classification. The Division of Mineral Classification is divided into four sections—coal, oil, phosphate and metalliferous—a committee, with a chairman in charge, being organized for each section. The chief of the division, who is directly responsible to the chief of the branch, is also the geologist in charge of the Metalliferous Section.

Division of Hydrographic Classification. The Division of Hydrographic Classification is composed of the Section of Utilization, the chairman of which is a hydraulic engineer, and the Section of Classification, the chairman of which is a geologist. As indicated by its title, this division classifies public lands with relation to their water resources.

Publication Branch. The Publication Branch edits, prepares for printing and engraving, and reads the proof of all publications—including text, maps and illustrations—issued by the Survey. It operates the Survey's engraving and printing plant and distributes all the publications of the Survey. It also develops, prints and prepares either for publication or for filing all photographs taken by members of the Survey.

The branch is organized in four divisions, each in charge of a chief who reports directly to the Director, namely:

1. Division of Book Publication
2. Division of Map Editing
3. Division of Engraving and Printing
4. Division of Distribution

Of the 934 permanent employees of the Survey on June 30, 1917, 174 constituted the personnel of the Publication Branch.

Division of Book Publication. The Division of Book Publication, in charge of an editor, includes the section of texts, in which all manuscripts are edited and all proofs are read, and the section of illustrations, in which drawings and engravings used to illustrate official publications are prepared for this purpose.

Division of Map Editing. The Division of Map Editing is charged with the maintenance of the accepted Survey standard in all maps, geologic and topographic, issued by the Survey. It includes a section of geologic maps and a section of topographic maps. These sections are regarded as a part of the Publication Branch, but their work is so intimately associated with the work performed in the Geologic and Topographic Branches that they are in fact administered as parts of those branches.

Division of Engraving and Printing. The Division of Engraving and Printing, administered by the chief engraver, is charged with all mechanical work involved in the reproduction of official maps or other illustrations published directly by the Survey. It includes an engraving and printing plant, an instrument shop and a photographic laboratory. It has been organized on a scale that permits it not only to do the work of this character required by the Survey but also to obtain from the Public Printer, by bidding in competition with private firms, work for other branches of the government.

Plant. The headquarters of the United States Geological Survey are in the Interior Department building, on the block bounded by E and F and 18th and 19th Streets, N. W., at Washington.

The chemical laboratory, in which is performed the analytical work necessary for the completion of geologic field investigations and some other researches in chemical geology, has equipment valued at $22,000. The equipment in the physical laboratory is valued at $5,000. The equipment used in the engraving and printing plant is valued at over $1,000,000. This equipment includes 10 power presses and 13 hand presses. The library contains about 120,000 volumes, 100,000 pamphlets and 25,000 maps.

The Survey maintains branch offices at Denver, Salt Lake City, San Francisco and other places, most of them in government buildings.

APPENDIX 1

OUTLINE OF ORGANIZATION

Explanatory Note

The Outlines of Organization have for their purpose to make known in detail the organization and personnel possessed by the several services of the national government to which they relate. They have been prepared in accordance with the plan followed by the President's Commission on Economy and Efficiency in the preparation of its outlines of the organization of the United States government.[1] They differ from those outlines, however, in that while the commission's report showed only organization units, the presentation herein has been carried far enough to show the personnel embraced in each organization unit.

These outlines are of value not merely as an effective means of making known the organization of the several services. If kept revised to date by the services they constitute exceedingly important tools of administration. They permit the directing personnel to see at a glance the organization and personnel at their disposition. They establish definitely the line of administrative authority and enable each employee to know his place in the system. They furnish the *essential basis* for making plans for determining costs by organization division and subdivision. They afford the data for a consideration of the problem of classifying and standardizing personnel and compensation. Collectively, they make it possible to determine the number and location of organization divisions of any particular kind, as for example—laboratories, libraries, blueprint rooms, or any other kind of plant possessed by the na-

[1] House Doc. 458, 62d Cong., 2d Sess. 1912—2 vols.

tional government, to what services they are attached and where they are located, or to determine what services are maintaining stations at any city or point in the United States.

As regards individual positions, changes are of course constantly taking place, both as to character and compensation. This is of course especially true at the present time, owing to the abnormal conditions resulting from the war.

OUTLINE OF ORGANIZATION

U. S. GEOLOGICAL SURVEY

DEPARTMENT OF THE INTERIOR

March 1, 1917[1]

	Number	Salary Rate
1. General Administrative Branch		
1. Office proper of the Director		
Director	1	$6,000.00
Private Secretary to the Director	1	1,740.00
Clerk	1	1,600.00
Messenger	1	840.00
2. Office of Chief Clerk		
1. Office proper of Chief Clerk		
Chief Clerk	1	2,500.00
Messenger	1	600.00
2. Executive Division		
1. Office of Chief of Executive Division		
Chief of Executive Division	1	2,400.00
Junior Clerk	1	1,200.00
Messenger	1	480.00
2. Mails, records and files		
Clerk	1	1,620.00
"	1	1,560.00
Junior Clerk	1	1,440.00
" "	1	1,200.00
Under Clerk	1	1,080.00
" "	1	1,020.00
" "	1	1,000.00
" "	1	900.00
Copyist	1	720.00
3. Appointments		
Clerk	1	1,740.00
Under Clerk	1	1,140.00
Skilled Laborer	1	960.00

[1] The personnel of the several units is being constantly changed. As occasion demands, employees are transferred from one division to another.

	Number	Salary Rate
1. General Administrative Branch.—*Cont.*		
2. Office of Chief Clerk.—*Cont.*		
2. Executive Division.—*Cont.*		
4. Multigraph Room		
Clerk	1	$1,620.00
Skilled Laborer	1	1,200.00
Assistant Map Printer	1	900.00
Messenger	1	780.00
“	2	540.00
Apprentice	1	420.00
“ (per diem)	1	1.25
3. Office of Chief Disbursing Clerk		
Chief Disbursing Clerk	1	2,500.00
Cashier	1	2,400.00
Junior Clerk	1	1,400.00
4. Division of Accounts		
Senior Clerk	1	1,920.00
Accountant	1	1,800.00
Clerk	2	1,680.00
“	1	1,560.00
Junior Clerk	1	1,380.00
“ “	1	1,320.00
“ “	1	1,200.00
Under Clerk	1	1,080.00
“ “	2	1,020.00
Messenger	1	720.00
“	1	480.00
“ (per diem)	1	1.50
Assistant Map Printer (per diem)	1	1.25
5. Library Division		
Librarian	1	2,000.00
Assistant Librarian	1	1,680.00
Cataloguer	1	1,380.00
Stenographer and Typewriter	1	1,020.00
Skilled Laborer	1	1,020.00
Clerk	1	1,000.00
Laborer	1	900.00
6. Engineer, Messengers and Labor Force		
1. Engineer Force		
Assistant Electrical Engineer	1	1,800.00
Engineer	1	1,200.00
Expert Mechanic	1	1,200.00
Wireman	1	1,020.00
Plumber and Steamfitter	1	840.00
Firemen	2	720.00
2. Carpenter Force		
Carpenter and Cabinet-Maker	1	1,200.00
Carpenter	1	900.00
Cabinet-Maker	1	900.00
3. Labor Force		
Laborer	1	780.00

	Number	Salary Rate
1. General Administrative Branch.—*Cont.*		
6. Engineer, Messengers and Labor Force.—*Cont.*		
3. Labor Force.—*Cont.*		
Laborer	1	$720.00
"	3	660.00
"	1	600.00
"	1	540.00
"	10	480.00
"	2	420.00
" (per month)	1	40.00
Janitor	1	600.00
Charwoman	1	270.00
" (per diem)	1	.....
4. Watch Force		
Lieutenant of the Watch	1	1,020.00
" " " "	2	840.00
Watchman	1	840.00
"	6	720.00
5. Messenger Force		
Messenger	1	840.00
"	2	420.00
"	6	360.00
Assistant Map Printer	1	360.00
" " " (per diem)	1	1.25
" " " (per diem)	6	1.00
Apprentice (per diem)	2	1.25
2. Topographic Branch		
1. Administrative and General		
Chief Geographer	1	4,000.00
Geographers (Computing)	2	3,000.00
" (Inspector)	1	2,500.00
Clerks	2	1,740.00
"	3	1,560.00
Junior Clerk	1	1,440.00
" "	1	1,200.00
Under Clerk	1	1,140.00
" "	1	1,020.00
" "	2	960.00
" "	1	840.00
Topographic Engineers (Computing)	3	2,520.00
" " (Computing)	1	2,340.00
" " (Compiling guidebooks)	1	2,340.00
" "	1	2,160.00
Topographer (Computing)	1	2,000.00
"	1	1,800.00
Assistant Topographer	1	1,740.00
" " (Inspector)	1	1,620.00
Draftsman	1	2,520.00
"	1	1,800.00

		Number	Salary Rate
2. Topographic Branch.—*Cont.*			
1. Administrative and General.—*Cont.*			
Draftsman		1	$1,620.00
Topographic Draftsman		2	1,500.00
" "		2	1,380.00
Skilled Mechanician		1	1,260.00
Laborer		1	780.00
2. Atlantic Division			
Geographer		1	3,600.00
Topographic Engineer		1	2,520.00
" "		1	2,460.00
" "		2	2,340.00
" "		3	2,280.00
" "		2	2,160.00
" "		2	2,000.00
" "		2	1,920.00
Topographer		1	2,000.00
"		1	1,800.00
"		1	1,740.00
Assistant Topographer		1	1,740.00
" "		3	1,620.00
" "	(per month)	1	145.00
" "	(per month)	2	135.00
" "	(per month)	1	115.00
" "	(per month)	6	100.00
Junior Topographer	(per month)	1	85.00
" "	(per month)	1	75.00
" "	(per month)	3	70.00
3. Central Division			
Geographer		1	3,600.00
Topographic Engineer		1	3,000.00
" "		1	2,280.00
" "		3	2,160.00
" "		4	1,920.00
Topographers		2	1,800.00
Assistant Topographer		1	1,380.00
" "	(per month)	1	145.00
" "	(per month)	2	135.00
" "	(per month)	2	125.00
" "	(per month)	1	100.00
Junior Topographers	(per month)	4	85.00
" "	(per month)	1	80.00
" "	(per month)	3	75.00
" "	(per month)	2	65.00
4. Rocky Mountain Division			
Geographer		1	2,700.00
Topographic Engineer		1	2,280.00
" "		1	1,980.00
" "		1	1,920.00
" "	(per month)	1	180.00
" "	(per month)	1	165.00

		Number	Salary Rate
2. Topographic Branch.—*Cont.*			
4. Rocky Mountain Division.—*Cont.*			
Topographer		I	$1,800.00
Assistant Topographer		I	1,740.00
" "		I	1,620.00
" "	(per month)	3	135.00
" "	(per month)	3	125.00
" "	(per month)	2	115.00
" "	(per month)	4	100.00
Junior Topographers	(per month)	3	85.00
" "	(per month)	2	75.00
5. Northwestern Division			
Geographer		I	3,600.00
Topographic Engineer		I	1,980.00
Topographers		2	1,800.00
Assistant Topographer		I	1,800.00
Junior Topographer	(per month)	I	90.00
6. Pacific Division			
Geographer		I	3,300.00
Topographic Engineer		I	2,700.00
" "		I	2,160.00
" "	(per month)	I	180.00
Topographers		2	1,800.00
"	(per month)	2	150.00
Assistant Topographer		I	1,740.00
" "	(per month)	I	135.00
" "	(per month)	3	115.00
" "	(per month)	3	100.00
Junior Topographer	(per month)	I	85.00
" "	(per month)	I	65.00
7. Division of Military Surveys			
Topographic Engineer		I	2,700.00

In addition to the above, the following (belonging to the Topographic Branch) are on furlough or temporarily with other Services:

Junior Topographer (Michigan State Highway Commission)	(per month)	I	80.00
Junior Topographer (Coast and Geodetic Survey)	(per month)	I	75.00
Topographer (Alaskan Engineering Commission)	(per diem)	I	5.00
Topographic Engineer (furlough)	(per diem)	I	7.50
Junior Topographic (furlough)	(per month)	I	85.00
" " (furlough)	(per month)	I	80.00
" " (furlough)	(per month)	I	70.00

		Number	Salary Rate
3. Geologic Branch			
1. Division of Geology			
1. Administrative and General			
Chief Geologist (Part of time in Section of the Geology of Oil and Gas Fields)		1	$4,500.00
Geologists	(per diem)	2	13.00
Senior Clerks		2	1,980.00
" "		1	1,920.00
Geologic Draftsman	Employed in this Section, but belonging under Section of illustrations under the Publication Branch)	1	1,800.00
Draftsman		1	1,740.00
Topographic Draftsman		1	1,680.00
Copy. Topographic Draftsman		1	1,200.00
Apprentice Draftsman		1	1,000.00
Copyist		1	720.00
Clerks		6	1,500.00
"		2	1,200.00
"	(per diem)	1	4.00
Junior Clerk		1	1,500.00
" "		1	1,440.00
" "		3	1,320.00
" "		3	1,200.00
" "		1	1,140.00
" "	(per diem)	1	4.75
Under Clerks		2	1,080.00
" "		2	1,020.00
" "		2	900.00
Library Assistant		1	1,320.00
Skilled Laborer (Prep. of Rock Specialist)		1	720.00
Apprentice		1	480.00
2. Section of Eastern Areal Geology			
Geologist (Chief of Section)		1	4,000.00
" (Part of time only. Chief of Section of Geologic Map Editing under Publication Branch)		1	3,000.00
Geologists		2	2,400.00
"		1	2,280.00
"	(per diem)	2	7.00
Associate Geologist		1	1,980.00
" "		1	1,860.00
Geologic Aid	(per diem)	1	4.00
3. Section of Western Areal Geology			
Geologist (Chief of Section)		1	2,700.00
"		2	2,700.00
"		1	2,280.00
"	(per diem)	1	12.00

	Number	Salary Rate
3. Geologic Branch.—*Cont.*		
1. Division of Geology.—*Cont.*		
3. Section of Western Areal Geology.—*Cont.*		
Geologist (per diem)	1	$ 7.00
Associate Geologist	1	2,160.00
" "	1	1,740.00
4. Section of Coastal Plain Investigations		
Geologist (Chief of Section; part time in Section of Paleontology and Stratigraphy)	1	3,300.00
Geologist (Part time in Section of the Geology of Oil and Gas Fields)	1	3,000.00
Geologist	1	2,700.00
Associate Geologist	1	1,620.00
" " (Part time in Section of Paleontology and Stratigraphy)	1	1,620.00
Assistant Geologist (Part time in Section of Paleontology and Stratigraphy) (per diem)	1	6.50
5. Section of Glacial Geology		
Geologist (Chief of Section)	1	2,700.00
"	1	2,700.00
6. Section of Paleontology and Stratigraphy		
Geologist (Chief of Section)	1	3,600.00
" (See under Section of Coastal Plain Investigations)	1	....
Geologist	1	3,000.00
"	2	2,880.00
Associate Geologist } See under Section of Coastal Plain Investigations	1	....
Assistant Geologist } See under Section of Coastal Plain Investigations	1	....
Geologist and Paleontologist	1	2,400.00
Paleontologist	1	2,000.00
Preparator of Fossils	1	1,500.00
" " "	1	1,380.00
" " "	1	1,260.00
" " "	1	1,140.00
" " "	1	1,020.00
Preparator (per hour)	1	.20
7. Section of the Geology of Metalliferous Deposits		
Geologist (Chief of Section)	1	4,500.00
"	2	3,300.00
" (Part time in Division of Mineral Resources)	4	3,300.00
Geologists	2	3,000.00
" (In Charge of Subsec-		

	Number	Salary Rate
3. Geologic Branch.—*Cont.*		
1. Division of Geology.—*Cont.*		
7. Section of the Geology of Metalliferous Deposits.—*Cont.*		
tion of Investigations in Petrology)	1	$2,700.00
Geologist	1	2,700.00
"	1	2,400.00
"	1	2,220.00
Associate Geologist (Part time in Division of Mineral Resources)	1	2,160.00
Associate Geologist	1	1,740.00
Assistant Geologist	1	2,100.00
8. Section of the Geology of Nonmetalliferous Deposits		
Geologist (Chief of Section)	1	3,300.00
"	1	3,000.00
"	1	2,700.00
" (per diem)	1	7.00
Associate Geologist	1	2,160.00
" "	1	2,100.00
9. Section of the Geology of Eastern Coal Fields		
Geologist (Chief of Section)	1	4,000.00
"	1	2,700.00
Geologic Aid (per month)	1	75.00
10. Section of the Geology of Western Coal Fields		
Geologist (Chief of Section)	1	4,000.00
"	1	2,700.00
" (per diem)	1	7.00
Associate Geologist	1	2,100.00
Assistant Geologist	1	1,380.00
" "	1	1,320.00
11. Section of the Geology of Oil and Gas Fields		
Chief Geologist (Chief of Section) (See under Administrative and General Section)	1	....
Geologist (See under Section of Coastal Plain Investigation)	1	....
Geologist	1	3,000.00
"	1	2,580.00
"	2	2,220.00
"	1	2,160.00
Associate Geologist	1	2,100.00
" "	4	1,980.00
" "	1	1,960.00
" "	1	1,860.00
" "	1	1,740.00

	Number	Salary Rate
3. Geologic Branch.—*Cont.*		
1. Division of Geology.—*Cont.*		
11. Section of the Geology of Oil and Gas Fields.—*Cont.*		
Associate Geologist	1	$1,680.00
Assistant Geologist	1	1,500.00
" "	1	1,320.00
Paleontologist	1	2,000.00
2. Division of Chemical and Physical Research		
1. Administrative and General		
Geologist (Chief of Division)	1	4,000.00
Junior Clerk	1	1,320.00
Skilled Laborer	1	1,080.00
Apprentice	1	420.00
2. Section of Chemistry		
Chemist	1	3,000.00
"	1	2,700.00
"	1	2,400.00
Assistant Chemists	2	1,980.00
" "	1	1,560.00
Junior Chemist	1	1,320.00
Machinist (per diem)	1	3.75
3. Section of Physics		
Physical Chemist	1	2,700.00
Physical Geologist	1	2,580.00
Assistant Physical Geologist	1	1,620.00
3. Division of Mineral Resources		
1. Administrative and General		
Geologist (Chief of Division)	1	3,600.00
Statistician	1	2,100.00
Statistical Experts	2	1,440.00
Assistant Statistician	1	1,440.00
Clerk	1	1,400.00
Junior Clerks	6	1,440.00
" "	1	1,380.00
" "	2	1,260.00
" "	2	1,200.00
Under Clerks	2	1,140.00
" "	4	1,080.00
" "	1	1,020.00
" "	3	960.00
" "	1	900.00
Skilled Laborer	1	1,200.00
Copyist	1	840.00
Messenger	1	480.00
2. Section of Metalliferous Deposits		
Geologists } See under Section of the Geology of Metalliferous Deposits, Division of Geology.	4	....
Associate Geologist }	1	....

	Number	Salary Rate
3. Geologic Branch.—*Cont.*		
3. Division of Mineral Resources.—*Cont.*		
3. Section of Non-metalliferous Deposits		
Geologists	2	$2,400.00
4. Field Offices (Denver, Salt Lake City and San Francisco)		
Statistician	1	3,000.00
"	1	2,700.00
"	1	2,400.00
"	1	2,280.00
"	1	1,740.00
Junior Clerk	1	1,380.00
Under Clerk	1	900.00
" " (per diem)	1	3.50
4. Division of Alaskan Mineral Resources		
1. Administrative and General		
Chief Geologist (Chief of Division)	1	4,500.00
Senior Clerk	1	1,980.00
Clerk	1	1,560.00
Under Clerk	1	1,020.00
2. Technical Investigations and Mapping		
Geologist and Administrative Assistant (Acts as Director in latter's absence)	1	3,600.00
Geologist	1	3,000.00
"	1	2,700.00
"	2	2,400.00
Associate Geologists	5	1,980.00
Assistant Geologist	1	1,340.00
" "	1	1,200.00
Topographic Engineer	1	3,000.00
" "	2	2,340.00
" "	1	2,100.00

In addition to the above, the following are employed occasionally by the Geologic Branch:

Geologist	(per diem)	1	$10.00
"	(per diem)	1	8.00
"	(per diem)	11	7.00
Associate Geologist	(per diem)	1	6.00
Assistant Geologists	(per annum)	2	1,200.00
" "	(per month)	1	100.00
" "	(per diem)	1	7.00
" "	(per diem)	6	6.00
" "	(per diem)	7	5.00
" "	(per diem)	1	4.50
" "	(per diem)	1	4.10
" "	(per diem)	2	4.00
Geologic Aids	(per month)	3	90.00
" "	(per month)	2	85.00

3. Geologic Branch.—*Cont.*
   4. Division of Alaskan Mineral Resources.—*Cont.*
      2. Technical Investigations and Mapping.—*Cont.*

		Number	Salary Rate
Geologic Aids	(per month)	4	$80.00
" "	(per month)	2	75.00
" "	(per month)	1	65.00
" "	(per diem)	1	4.00
" "	(per diem)	1	3.50
Junior Geologist	(per month)	1	75.00
" "	(per diem)	2	4.00
" "	(per diem)	3	3.50
Geologist and Paleontologist	(per diem)	1	13.00
" " "	(per diem)	1	7.00
Assistant Topographer	(per diem)	1	4.00
Clerk	(per diem)	1	5.00
Stenographer	(per hour)	1	.50
"	(per hour)	1	.40

Standing Committees of the Geologic Branch:—
1. Committee on Geologic Names
   a. Chairman (Geologist—Chief of Section of Paleontology and Stratigraphy under Division Geology)
   b. Members
      1. Geologist (Chief of Section of Geologic Map Editing under Publication Branch)
      2. Geologist (In Section of Western Areal Geology under Division of Geology)
      3. Geologist (Chief of Section of Western Areal Geology under Division of Geology)
      4. Geologist (In Section of Coast Plain Investigations under Divison of Geology)
   c. Associate Members
      1. Chiefs of Section in Division of Geology
      2. Chief of Division of Alaskan Mineral Resources
2. Committee on Physiography
   a. Chairman (Geologist—Chief of Section of the Geology of Western Coal Fields under the Division of Geology)
   b. Members
      1. Geologist (Water Resources Branch)
      2. Geologist (In Section of Western Areal Geology under Division of Geology)
      3. Geologist (In Section of the Geology of Oil and Gas Fields under Division of Geology)
      4. Geologist and Administrative Assistant (In Division of Alaskan Mineral Resources).

4. Water Resources Branch
   1. Administrative and General

	Number	Salary Rate
Chief Hydraulic Engineer (Chief of Branch—also in charge of Division of Water Utilization)	1	$4,000.00

	Number	Salary Rate
4. Water Resources Branch—*Cont.*		
1. Administrative and General.—*Cont.*		
Senior Clerk	1	$1,920.00
Junior Clerk (Part time in Division of Ground Waters)	1	1,320.00
Under Clerk	1	1,080.00
Copyist	1	720.00
Messenger	1	480.00
2. Division of Surface Waters		
1. Middle Atlantic District (Office, Washington, D. C.)		
Hydraulic Engineer (In Charge of Division)	1	3,600.00
Engineer	1	2,160.00
Assistant Engineer	1	1,920.00
" "	2	1,620.00
Junior Engineer	1	1,440.00
" "	1	1,320.00
Senior Clerk (per diem)	1	6.00
Junior Clerk	1	1,320.00
Under Clerk	1	720.00
Stenographic Clerk (per diem)	1	3.00
2. New England District (Office: Boston, Mass.)		
Engineer	1	2,160.00
Junior Engineer	1	1,200.00
Under Clerk	1	900.00
3. New York District (Office: Albany, N. Y.)		
Hydraulic Engineer	1	3,000.00
Assistant Engineer	1	1,800.00
Junior Engineer	1	1,320.00
" "	1	1,200.00
Under Clerk	1	1,080.00
4. South Atlantic and Eastern Gulf District (Office: Atlanta, Ga.)		
Assistant Engineer	1	2,100.00
Under Clerk	1	900.00
5. Texas District (Austin, Texas)		
Engineer	1	2,160.00
Under Clerk	1	900.00
6. Ohio Valley District (Office: Temporarily at Washington, D. C.)		
Hydraulic Engineer	1	3,000.00
Assistant Engineer	1	1,740.00
Under Clerk	1	840.00
7. Upper Mississippi Valley and Great Lakes District. (Office: Madison, Wis.; Sub-Office: St. Paul, Minn.)		
Hydraulic Engineer (per diem)	1	8.00

	Number	Salary Rate
4. Water Resources Branch—*Cont.*		
2. Division of Surface Waters.—*Cont.*		
7. Upper Mississippi Valley and Great Lakes District. (Office: Madison, Wis.; Sub-Office: St. Paul, Minn.)—*Cont.*		
Assistant Engineer	1	$2,100.00
" "	1	1,500.00
Junior Engineer	1	1,200.00
" "	1	1,080.00
" " (per diem)	1	4.00
8. Upper Missouri District (Office: Helena, Mont.)		
Hydraulic Engineer	1	2,280.00
Assistant Engineer	1	1,620.00
" " (per diem)	1	7.00
Junior Engineer	1	1,320.00
Clerk	1	900.00
9. Rocky Mountain District (Office: Denver, Colo.)		
Hydraulic Engineer	1	3,000.00
Junior Engineer	1	1,320.00
" "	1	1,200.00
" "	1	1,080.00
" " (per diem)	1	5.00
Under Clerk	1	1,140.00
10. Idaho District (Office: Boise, Idaho)		
Assistant Engineer	1	2,100.00
Junior Engineer	1	1,200.00
Under Clerk	1	1,140.00
11. Great Basin District (Office: Salt Lake City, Utah)		
Assistant Engineer	1	1,860.00
" "	1	1,740.00
Junior Engineer	1	1,380.00
" "	1	1,320.00
Under Clerk	1	960.00
12. Washington District (Office: Tacoma, Wash.)		
Engineer	1	2,160.00
Junior Engineer	1	1,500.00
" "	1	1,440.00
" " (per diem)	1	4.00
Under Clerk	1	900.00
13. Oregon District (Office: Portland, Oregon)		
Hydraulic Engineer	1	2,700.00
Junior Engineer	1	1,440.00
" "	1	1,080.00
Junior Clerk	1	1,200.00

	Number	Salary Rate
4. Water Resources Branch—*Cont.*		
2. Division of Surface Waters.—*Cont.*		
14. California District (Office: San Francisco, Cal.; Sub-Office: Los Angeles, Cal.)		
Hydraulic Engineer	1	$2,700.00
Assistant Engineer	1	1,860.00
" "	1	1,500.00
" "	1	1,440.00
Junior Engineer	1	1,320.00
" "	2	1,080.00
Junior Clerk	1	1,320.00
Under Clerk (per diem)	1	3.25
15. Arizona District (Phoenix, Ariz.)		
Assistant Engineer	1	2,160.00
Junior Engineer	1	1,200.00
Under Clerk	1	900.00
16. Hawaiian District (Office: Honolulu, Hawaii)		
Hydraulic Engineer (Paid by Hawaii; nominal Salary of $12 per annum paid by U. S.)	1	12.00
Assistant Engineers	2	1,800.00
Junior Engineer	1	1,500.00
3. Division of Ground Waters		
Geologist (Chief of Division)	1	2,580.00
Geologist	1	2,400.00
Associate Engineer	1	1,680.00
Assistant Chemist	1	1,620.00
Assistant Geologist	1	1,500.00
" "	1	1,320.00
" "	1	1,200.00
Geologic Aid (per diem)	1	3.25
Junior Clerk	1	1,200.00
" " (See under Administrative and General)	1	....
4. Division of Water Utilization		
Chief Hydraulic Engineer (See under Administrative and General)	1	....
Hydraulic Engineer	1	2,880.00
" "	1	2,700.00
Junior Engineer	1	1,200.00
Junior Clerk	1	1,200.00
Draftsman (per hour)	1	.50
Stenographer (per hour)	1	.40

In addition to the above the following have been temporarily assigned to the Division of Alaskan Mineral Resources under the Geologic Branch, or are unassigned:

	Number	Salary Rate
4. Water Resources Branch—*Cont.*		
4. Division of Water Utilization.—*Cont.*		
Assistant Engineer (Alaskan Mineral Resources)	1	$1,800.00
Assistant Engineer (Alaskan Mineral Resources)	1	1,620.00
Assistant Engineer (Unassigned) (per diem)	1	7.00
5. Land Classification Board		
1. Administrative and General		
Geologist (Chief of Branch)	1	$4,000.00
Senior Clerk	1	1,800.00
Junior Clerks	1	1,320.00
" "	1	1,260.00
" "	2	1,200.00
" " (per diem)	1	5.00
Under Clerk	3	1,140.00
" "	3	1,080.00
" "	1	1,020.00
" "	1	900.00
" "	2	840.00
" "	1	780.00
Draftsman	1	1,560.00
Copy. Topographic Draftsman	1	1,260.00
" " "	1	1,200.00
Skilled Laborer	1	540.00
Laborer	1	900.00
Messenger	1	600.00
"	1	480.00
2. Division of Mineral Classification		
1. Coal Section		
Junior Topographer	1	1,140.00
2. Oil Section		
Assistant Geologist	1	1,500.00
3. Phosphate Section		
Associate Geologist	1	1,800.00
4. Metalliferous Section		
Geologist (Chief of Division)	1	2,700.00
3. Division of Hydrographic Classification		
1. Section of Utilization		
Hydraulic Engineer (Chairman of Section)	1	3,000.00
Assistant Engineer	1	1,860.00
" "	1	1,620.00
2. Section of Classification		
Geologist (Chairman of Section)	1	2,700.00
Assistant Classifier	1	1,440.00
Junior Engineers	2	1,380.00
Junior Topographer	1	1,020.00

	Number	Salary Rate
6. Publication Branch		
1. Division of Book Publication		
1. Texts Section		
Editor	1	$2,400.00
Assistant Editor	1	2,400.00
Senior Clerk	1	1,800.00
Copy Editor	1	1,740.00
Editorial Clerk	1	1,800.00
" "	1	1,560.00
" "	1	1,440.00
" " (per diem)	1	5.00
Assistant Map Printer	1	360.00
2. Illustrations Section		
Draftsman	1	2,400.00
"	1	1,920.00
"	2	1,740.00
Paleontologist Draftsman	1	1,800.00
" "	1	1,560.00
" "	1	1,500.00
" "	1	1,200.00
Artist Draftsman	1	1,200.00
Copy Topographic Draftsman	1	960.00
Clerk	1	900.00
Copyist	1	720.00
Messenger	1	480.00
3. Distribution Section		
Senior Clerk	1	2,100.00
Clerk	1	2,000.00
"	1	1,740.00
"	1	1,200.00
"	2	900.00
Junior Clerk	1	1,200.00
Under Clerk	1	1,140.00
" "	1	1,080.00
" "	1	1,000.00
" "	1	960.00
" "	3	900.00
" "	1	840.00
Skilled Laborer	1	960.00
" "	2	840.00
Laborer	1	960.00
"	1	600.00
Messengers	3	540.00
2. Division of Map Editing		
1. Section of Geologic Maps		
Editor (paid by Geologic Branch)	1	....
Map Reviser	1	1,620.00
Geologic Draftsman	1	1,980.00
Clerk	1	1,620.00
2. Section of Topographic Maps		
Editor	1	2,520.00

	Number	Salary Rate
6. Publication Branch.—*Cont.*		
2. Division of Map Editing.—*Cont.*		
2. Section of Topographic Maps.—*Cont.*		
Map Reviser	1	$1,620.00
" "	1	1,440.00
Topographic Draftsman	1	1,740.00
Clerk	1	1,620.00
3. Division of Engraving and Printing		
1. Engraving and Printing Section		
Chief Engraver	1	4,000.00
Senior Clerk	1	1,920.00
Under Clerk	1	1,080.00
Cartographic Specialist	1	1,620.00
Engraver	1	2,100.00
"	1	1,740.00
"	1	1,620.00
Copper-plate Map Engraver	2	1,920.00
" " "	1	1,800.00
" " "	1	1,620.00
" " "	3	1,560.00
" " "	2	1,440.00
" " "	1	1,400.00
" " "	1	1,260.00
" " "	5	1,200.00
Assistant Engravers	2	780.00
Assistant Map Engravers	4	720.00
" " "	3	660.00
Apprentice Engravers	2	660.00
Lithographic Map Engraver	1	2,100.00
" " "	2	1,680.00
" " "	1	1,500.00
" " "	2	1,260.00
" " "	1	720.00
Draftsman	1	1,200.00
Photolithographer	1	1,200.00
Wet-plate Photographer	1	1,380.00
Negative Cutter	2	1,200.00
" "	1	1,140.00
" "	1	1,020.00
" "	1	780.00
" "	1	720.00
Negative Cutter and Lithographic Draftsman	1	780.00
Printers	2	1,440.00
Map Printer	1	1,800.00
" "	1	1,620.00
" "	1	1,500.00
" "	2	1,320.00
" "	1	1,320.00
" "	2	1,260.00
" "	1	1,200.00

	Number	Salary Rate
6. Publication Branch.—*Cont.*		
3. Division of Engraving and Printing.—*Cont.*		
1. Engraving and Printing Section.—*Cont.*		
Map Printer	1	$1,020.00
" "	1	960.00
" " (per diem)	1	4.00
Assistant Map Printers	2	1,140.00
" " "	1	1,100.00
" " "	2	1,080.00
" " "	2	900.00
" " "	1	840.00
" " "	1	780.00
" " "	1	720.00
" " "	1	660.00
" " " (per diem)	5	2.00
" " " (per diem)	3	1.75
" " " (per diem)	6	1.50
" " " (per diem)	3	1.25
" " " (per diem)	7	1.00
Laborer	2	720.00
"	2	660.00
"	1	600.00
"	1	540.00
" (per diem)	1	1.50
Messenger	1	540.00
2. Instrument Shop		
Expert Mechanician	1	2,700.00
Skilled Laborer	1	720.00
" "	1	660.00
3. Photographic Laboratory		
Photographers	2	2,000.00
Assistant Photographers	2	1,620.00
" "	1	1,320.00
" "	2	1,200.00
" "	1	900.00
" "	1	720.00

APPENDIX 2

CLASSIFICATION OF ACTIVITIES

Explanatory Note

The Classifications of Activities have for their purpose to list and classify in all practicable detail the specific activities engaged in by the several services of the national government. Such statements are of value from a number of standpoints. They furnish, in the first place, the most effective showing that can be made in brief compass of the character of work performed by the service to which they relate. Secondly, they lay the basis for a system of accounting and reporting that will permit of the showing of total expenditures classified according to activities. Finally, taken collectively, they make possible the preparation of a general or consolidated statement of the activities of the government as a whole. Such a statement will reveal in detail, not only what the government is doing, but the services in which the work is being performed. For example, one class of activities that would probably appear in such a classification is that of "scientific research." A subhead under this class would be "chemical research." Under this head would appear the specific lines of investigation under way and the services in which they were being prosecuted. It is hardly necessary to point out the value of such information in planning for future work and in considering the problem of the better distribution and coordination of the work of the government. The Institute has it in contemplation to attempt such a general listing and classification of the activities of the government upon the completion of the present series.

Classification of Activities

1. Topographic Survey
2. Geologic Survey
3. Investigation of Mineral Resources
    1. Metalliferous Deposits
    2. Non-metalliferous Deposits
    3. Coal
    4. Oil and gas
4. Investigation of Water Resources
5. Classification of Public Lands
6. Compilation of Statistics
    1. Mineral Resources
    2. Production of Mines and Quarries
    3. Imports and Exports
7. Scientific Research
    1. Chemical
    2. Physical
    3. Geological
    4. Methods of Flood Prevention and Drainage
    5. Stream Pollution and Remedial Methods
8. Publication
    1. Topographic Maps
    2. Geologic Maps and Descriptive Text
    3. Description of the Mineral Resources of the United States
    4. Papers on Geological and Related Subjects
    5. Papers on Water Resources
    6. Miscellaneous
9. Operation of Engraving and Printing Plant

APPENDIX 3

PUBLICATIONS

Topographic Maps and Folios and Geologic Folios

The topographic maps and folios and the geologic folios issued by the United States Geological Survey may be obtained from the Director of the United States Geological Survey, Washington, D. C.

A large pamphlet, issued by the Survey, entitled "Topographic Maps and Folios and Geologic Folios" contains a list of all the topographic maps and folios and the geologic folios. A list of the geologic folios is also contained in a Survey publication entitled "Publications of the United States Geological Survey." The Survey publishes a set of "index circulars" which show the maps and folios completed for different regions of the country and the stages of the work in areas for which maps and folios are in preparation. In addition the Survey has prepared a circular which gives a list of the maps which illustrate various physiographic types and a general circular in regard to the geologic folios.

Topographic Maps and Folios

The topographic maps and folios published by the Survey fall into three classes:

1. Topographic Atlas Sheets
2. Topographic Folios
3. Special Topographic Maps

Topographic Atlas Sheets. Under this title are published by far the greatest number of topographic maps. The price of each standard atlas sheet is 10 cents.

Topographic Folios. Three topographic folios, consisting of separate topographic atlas sheets, assembled to illustrate types of topography, have been published. No others are projected. The price of two of these folios is 25 cents each. The price of the other folio is 50 cents.

Special Topographic Maps. These usually consist of several topographic atlas sheets or parts of sheets combined on a single sheet, sometimes on a larger or smaller scale than the regular sheets.

Geologic Folios

The geologic folios published by the Survey fall into three classes

1. Geologic Atlas Folios
2. Geologic Atlas Sheets
3. Special Geologic Maps

Geologic Atlas Folios. Under this title data regarding geologic conditions is presented. A folio contains both maps and descriptive matter. The price of folios ranges from 25 cents a copy to $1.00 a copy, except that certain folios damaged by fire and water in 1913 are obtainable at 5 cents a copy.

Geologic Atlas Sheets. Separate sheets forming a part of the geologic folios are issued under this title in order to avoid the necessity for supplying an entire folio when one or two sheets will meet the requirements of persons desiring certain special data. The price of each sheet is 10 cents.

Special Geologic Maps. Only a few maps of this kind have been issued. They are prepared in order to present conditions at some particular place or in a way different from that employed in geologic folios. The prices of the several special maps are listed in the pamphlets and index circulars above mentioned.

Other Publications

The remaining publications of the Survey, with the exception of Monographs, may be obtained free from the Director of the United States Geological Survey as long as the Survey's supply lasts. It is possible then in some cases to purchase such publications from the Superintendent of Documents, Government Printing Office, Washington, D. C. All Monographs are obtainable only from the Superintendent of Public Documents and at prices stated.

These publications of the Survey are listed in a large pamphlet entitled, "The Publications of the U. S. Geological Survey," the latest volume of which was issued in December, 1916. This pamphlet is supplemented from time to time by the issue of circular lists of new publications. These publications are:

- Mineral Resources of the United States
- Technical treatises on geologic and related scientific subjects
  - Monographs
  - Professional Papers
  - Bulletins
- Water Supply Papers
- Annual Report of Director
- Miscellaneous

Mineral Resources of the United States. Each year the Survey issues one or two volumes entitled "Mineral Resources of the United States." These publications are, on request, distributed gratuitously as long as the Survey's stock lasts.

Technical Treatises on Geological and Related Scientific Subjects. These publications are classified as Bulletins, Professional Papers, and Monographs, the line of demarcation being based upon the extensiveness and exhaustiveness of the work. The Bulletins and Professional Papers are distributed gratuitously as long as the Survey's stock lasts. The Monographs are sold at prices ranging from eighty cents up.

Reference has been made in Chapter II to the tourist guide-books of the western United States issued for the first time in 1915 and listed under the title of Bulletins. Like other Bulletins these guide-books are distributed gratuitously as long as the Survey's supply lasts.

Water Supply Papers. A number of these papers are published each year. They may be obtained free until the Survey's stock is exhausted.

Annual Report of the Director. The annual report of the Director gives general information regarding the operations of the Survey during the year to which it relates. It differs in no essential respect from the annual reports of other administrative officers.

Miscellaneous (Special Publications). The miscellaneous or special publications of the Survey consist of pamphlets or books relating to office or field administration, a press bulletin announcing the publication of the Survey's reports, and other matter designed to advertise them or to inform the public of their nature.

APPENDIX 4

LAWS

(A) Index to Laws

Creation

- Geological Survey established....................20 Stat. L., 394
- Other surveys discontinued ....20 Stat. L., 394

Personnel

- Positions established
  - Director ....20 Stat. L., 394
  - Acting Director ....28 Stat. L., 197
- Method of appointment of
  - Director ....20 Stat. L., 394
  - Acting Director ....28 Stat. L., 197
  - Scientific employees ....23 Stat. L., 194
- Qualifications of
  - Acting Director ....28 Stat. L., 197
  - Scientific employees ....23 Stat. L., 194
- Restrictions on
  - Director ....20 Stat. L., 394
  - Members ....20 Stat. L., 394
- Duties of
  - Director ....20 Stat. L., 394
  - Acting Director ....28 Stat. L., 197
- Salaries of
  - Director ....40 Stat. L., 144
  - Certain scientific assistants....................40 Stat. L., 144
  - Certain clerks, photographers, copyists, etc.....40 Stat. L., 144

Activities

- General administration
  - Pay of field employees, assignments may be permitted ....34 Stat. L., 727
  - Expenses of employees in the field, paid from personal funds, may be reimbursed........34 Stat. L., 727
  - Facilities for research and illustration to be accessible to scientific investigators, etc......27 Stat. L., 395
  - Collections of rocks, minerals, fossils, etc., to be deposited in the National Museum........20 Stat. L., 394
  - Sales of copies of material in possession of Survey authorized ....40 Stat. L., 144

Activities.—Cont.

General administration.—*Cont.*

- Cartographic or other engraved or lithographic data ... 34 Stat. L., 728
- Photographs or lantern slides ... 35 Stat. L., 989
- Making of purchase of supplies or procurement of services outside the District of Columbia, in open market permitted ... 39 Stat. L., 301

Topographic surveys

- Preference to be given to requests made by War Department ... 39 Stat. L., 301
- Base level to be determined and marked by posts or bench marks ... 29 Stat. L., 435
- Exterior boundaries of public forest reservations to be established and marked in certain cases ... 30 Stat. L., 1074

Publications[1]

- In general
  - Of what to consist ... 32 Stat. L., 741
  - Size of editions ... 32 Stat. L., 741
  - Transmission to Library of Congress of copies of ... 32 Stat. L., 741
- Specific publications
  - Specific estimates and appropriations required prior to publication
    - Scientific reports known as monographs and bulletins ... 24 Stat. L., 255
    - Engravings for annual reports or for monographs or bulletins ... 24 Stat. L., 255
    - Engravings of illustrations, sections and maps ... 24 Stat. L., 255
  - Size of page (of annual report of Director, monographs, professional papers, bulletins, mineral resources, water supply and irrigation papers) ... 32 Stat. L., 741
  - Number of volumes: annual report of the Director ... 32 Stat. L., 741
  - Gratuitous distribution of
    - To public libraries—
      - Special memoirs and reports ... 24 Stat. L., 647
      - Monographs, bulletins and reports ... 29 Stat. L., 465
      - Sale publications ... 32 Stat. L., 741
    - To foreign governments, literary and scientific associations, etc.—
      - Topographic and geologic maps and atlases ... 29 Stat. L., 701
    - Generally—bulletins and professional papers ... 32 Stat. L., 741

[1] The provisions of the acts of Mar. 3, 1879, and March 2, 1895 (20 Stat. L., 394 and 28 Stat. L., 960) relative to publications are not here indexed. See note on p. 113.

Items of appropriation[1]

- Salaries
  - Office of Director
    - Director ....40 Stat. L., 144
    - Chief Clerk ....40 Stat. L., 144
    - Chief disbursing clerk....................40 Stat. L., 144
    - Librarian ....40 Stat. L., 144
    - Photographer ....40 Stat. L., 144
    - Assistant photographers (2)....................40 Stat. L., 144
    - Clerks (5) ....40 Stat. L., 144
    - Copyists (4) ....40 Stat. L., 144
    - Watchmen (4) ....40 Stat. L., 144
    - Janitor (1) ....40 Stat. L., 144
    - Messenger boys (4) ....40 Stat. L., 144
  - Scientific assistants
    - Geologists (2)....................40 Stat. L., 144
    - Paleontologists (2) ....40 Stat. L., 144
    - Chemist (1) ....40 Stat. L., 144
    - Geographers (2) ....40 Stat. L., 144
    - Topographers (2) ....40 Stat. L., 144
  - Skilled laborers and various temporary employees ....40 Stat. L., 144
- Purchase of books, directories and scientific periodicals ....40 Stat. L., 144
- Specific activities
  - Topographic surveys
    - In various portions of the United States......40 Stat. L., 144
    - Of public lands designated or to be designated as national forests ....40 Stat. L., 144
  - Geologic surveys in various portions of the United States ....40 Stat. L., 144
  - Chemical and physical researches relating to geology ....40 Stat. L., 144
  - Examination and classification of lands for determination of suitability for enlarged homesteads, etc. ....40 Stat. L., 144
  - Preparation of report on mineral resources....40 Stat. L., 144
  - Investigation of mineral resources of Alaska....40 Stat. L., 144
  - Gauging streams and other water resources investigations, and preparation of reports....40 Stat. L., 144
  - Discovering, developing, etc., springs, streams, etc., on arid public lands, etc.............40 Stat. L., 144
  - Preparation of illustrations....................40 Stat. L., 144
  - Engraving of illustrations for annual report of Director and other publications..........40 Stat. L., 174
  - Engraving and printing
    - Geologic maps ....40 Stat. L., 174
    - Topographic data collected for War Department ....40 Stat. L., 174
  - Printing and binding of publications..........40 Stat. L., 174

[1] All the items under this head are found in the appropriation act for the fiscal year ending June 30, 1918, approved June 12, 1917.

(B) Compilation of Laws

1879—Act of March 3, 1879; 20 Stat. L., 394. (An act making appropriations for sundry civil expenses of the government for the fiscal year ending June thirtieth, eighteen hundred and eighty, and for other purposes.)

For the salary of the Director of the Geological Survey, which office is hereby established, under the Interior Department, who shall be appointed by the President by and with the advice and consent of the Senate, six thousand dollars: Provided, That this officer shall have the direction of the Geological Survey, and the classification of the public lands and examination of the geological structure, mineral resources and products of the national domain. And that the Director and members of the Geological Survey shall have no personal or private interests in the lands or mineral wealth of the region under survey, and shall execute no surveys or examinations for private parties or corporations; and the Geological and Geographical Survey of the Territories, and the Geographical and Geological Survey of the Rocky Mountain Region, under the Department of the Interior, and the Geographical Surveys West of the One-hundredth Meridian, under the War Department, are hereby discontinued, to take effect on the thirtieth day of June, eighteen hundred and seventy-nine. And all collections of rocks, minerals, soils, fossils, and objects of natural history, archæology, and ethnology, made by the Coast and Interior Survey, the Geological Survey, or by any other parties for the Government of the United States, when no longer needed for investigations in progress shall be deposited in the National Museum.[1]

1880—Act of June 16, 1880; 21 Stat. L., 274. (An act making appropriations for the sundry civil expenses of the government for the fiscal year ending June thirtieth, eighteen hundred and eighty-one, and for other purposes.)

And the Secretary of War is hereby authorized to detail not exceeding two officers of the Ordnance Corps to serve with the Geological Survey: Provided, That in his judgment it can be done without injury to the Service.

[1] Provision was further made in this act for the publication of an annual report of operations to "accompany the annual report of the Secretary of the Interior" and for the publication of special memoirs and reports to "be issued in uniform quarto series if deemed necessary by the Director, but otherwise in ordinary octavos." While not specifically repealed these provisions are doubtless to be regarded as superseded by the Joint Resolution of May 16, 1902 (32 Stat. L., 741).

1884—Act of July 5, 1884; 23 Stat. L., 111. (An act making appropriations for the support of the army for the fiscal year ending June thirtieth, eighteen hundred and eighty-five, and for other purposes.)

Provided also, That hereafter the Quartermaster-General and his officers, under his instructions, wherever stationed, shall receive, transport, and be responsible for all property turned over to them, or anyone of them, by the officers or agents of any Government survey, for the National Museum, or for the civil or naval departments of the Government, in Washington or elsewhere, under the regulations governing the transportation of Army supplies, the amount paid for such transportation to be refunded or paid by the Bureau to which such property or stores pertain.

1884—Act of July 7, 1884; 23 Stat. L., 194. (An act making appropriations for sundry civil expenses of the government for the fiscal year ending June thirtieth, eighteen hundred and eighty-five, and for other purposes.)

And the scientific employees of the Geological Survey shall be selected by the Director, subject to the approval of the Secretary of the Interior, exclusively for their qualifications as professional experts.

1886—Act of August 4, 1886; 24 Stat. L., 255. (An act making appropriations for sundry civil expenses of the government for the fiscal year ending June thirtieth, eighteen hundred and eighty-seven, and for other purposes.)

And hereafter the scientific reports known as the monographs and bulletins of the Geological Survey shall not be published until specific and detailed estimates are made therefor, and specific appropriations made in pursuance of such estimates; and no engraving for the annual reports or for such monographs and bulletins, or of illustrations, sections, and maps, shall be done until specific estimates are submitted therefor and specific appropriations made based on such estimates: Provided further, That all printing and engraving for the Geological Survey, and Coast and Geodetic Survey, the Hydrographic office of the Navy Department, and the Signal Service shall hereafter be estimated for separately and in detail, and appropriated for separately for each of said Bureaus.[1]

[1] The provisions of the above act from the words "the scientific reports known, etc.," to the words "based on such estimates" were reënacted in the act of January 12, 1895; 28 Stat. L., 621.

1887—Act of March 3, 1887; 24 Stat. L., 509. (An act making appropriations for sundry civil expenses of the government for the fiscal year ending June thirtieth, eighteen hundred and eighty-eight, and for other purposes.)

Hereafter the estimates for the Geological Survey shall be itemized.

1887—Act of March 3, 1887; 24 Stat. L., 647. (Joint resolution to distribute copies of special memoirs and reports of the United States Geological Survey.)

Resolved, etc., That there shall be distributed from the number of special memoirs and reports of the United States Geological Survey now authorized by law one copy of every such publication to every public library which shall be designated to the Secretary of the Interior as follows:

Two public libraries to be designated by each of the Senators from the States, respectively, two public libraries by the Representative in Congress from every Congressional district and two public libraries by the Delegate from every Territory; such public libraries to be additional to those to which the said publications are distributed under existing law.

1892—Act of April 12, 1892; 27 Stat. L., 395. (Joint resolution to encourage the establishment and endowment of institutions of learning at the national capital by defining the policy of the government with reference to the use of its literary and scientific collections by students.)

Resolved, by the Senate and House of Representative of the United States of America, in Congress assembled, That the facilities for research and illustration in the following and any other governmental collections now existing or hereafter to be established in the city of Washington for the promotion of knowledge shall be accessible, under such rules and restrictions as the officers in charge of each collection may prescribe, subject to such authority as is now or may hereafter be permitted by law, to the scientific investigators and to students of any institution of higher education now incorporated or hereafter to be incorporated under the laws of Congress or of the District of Columbia, to wit:

* * *

Of the Geological Survey.

1894—Act of July 31, 1894; 28 Stat. L., 197. (Legislative, Executive and Judicial Appropriation Act.)

The Secretary of the Interior may hereafter authorize one of the geologists to act as Director of the Geological Survey in the absence of that officer.

1896—Act of June 11, 1896[1]; 29 Stat. L., 435. (An act making appropriations for sundry civil expenses of the government for the fiscal year ending June thirtieth, eighteen hundred and ninety-seven, and for other purposes.)

United States Geological Survey. * * * For topographic surveys in various portions of the United States * * * Provided, That hereafter in such surveys west of the ninety-fifth meridian elevations above a base level located in each area under survey shall be determined and marked on the ground by iron or stone posts or permanent bench marks, at least two such posts or bench marks to be established in each township or equivalent area except in the forest-clad and mountain areas, where at least one shall be established, and these shall be placed, whenever practicable, near the township corners of the public land surveys; and in the areas east of the ninety-fifth meridian at least one such post or bench mark shall be similarly established in each area equivalent to the area of a township of the public land surveys. * * *

1896—Act of February 26, 1896; 29 Stat. L., 465. (Joint resolution extending the provisions of section seventy-nine of "An Act providing for the public printing and binding and the distribution of public docu-

[1] The Act of March 2, 1895, provided for the incorporation into the report of the Director of the report of the mineral resources of the United States, and made the further provision "and that the separate chapters on any given mineral product, such as iron, coal, building stone, and so forth, shall be printed as rapidly as transmitted for publication; that a pamphlet edition of any chapter shall be printed for distribution on the request of the Director of the Geological Survey, approved by the Secretary of the Interior, the size of the edition to be controlled by the importance of the mineral treated; that hereafter papers for the Director's annual report that are of a strictly economic character shall be issued in pamphlet form, in the same manner as prescribed above for the report on the mineral resources; that the entire cost of paper, printing, and binding of all of the above provided for pamphlets shall not exceed two thousand dollars." While not specifically repealed, these provisions are doubtless to be regarded as superseded by the Joint Resolution of May 16, 1902 (32 Stat. L., 741).

ments," approved January twelfth, eighteen hundred and ninety-five, so as to include monographs, bulletins, and reports of the Geological Survey published in eighteen hundred and ninety-four and succeeding years.)

Resolved, by the Senate and House of Representatives of the United States of America in Congress assembled, That the provisions of section seventy-nine of "An Act providing for the public printing and binding and the distribution of public documents," approved January twelfth, eighteen hundred and ninety-five, which section reads as follows: "There shall be distributed of monographs, bulletins, and reports of the United States Geological Survey, now in possession of said Survey, being publications prior to the year eighteen hundred and ninety-four, one copy of every such publication to every public library which shall be designated to the superintendent of documents, as follows: Two public libraries to be designated by each of the Senators from the States, respectively, two public libraries by the Representatives in Congress, from each Congressional district, and two public libraries by the delegate from each Territory; such public libraries to be additional to those to which said publications are distributed under existing law," shall be extended to the monographs, bulletins and reports of the Geological Survey which were published during the year eighteen hundred and ninety-four, and to those which have been published since that year, and to those which may be published in the future: Provided, That nothing herein contained shall be construed to interfere with the distribution of memoirs and reports, so far as the same is provided for by the joint resolution "To distribute copies of special memoirs and reports of the United States Geological Survey," approved March third eighteen hundred and eighty-seven.

1897—Act of February 18, 1897; 29 Stat. L., 701. (Joint resolution providing for the distribution of the maps and atlases of the United States Geological Survey.)

Resolved, etc., That the Director of the Geological Survey be, and is hereby authorized and directed, on the approval of the Secretary of the Interior, to dispose of the topographic and geologic maps and atlases of the United States Geological Survey, * * * at such prices and under such regulations as may from time to time be fixed by him and approved by the Secretary of the Interior;

And that a number of copies of each map or atlas, not exceeding five hundred, shall be distributed gratuitously among foreign governments and departments of our own government, to literary and scientific associations, and to such educational institutions or libraries as may be designated by the Director of the Survey and approved by the Secretary of the Interior. * * *

1899—Act of March 3, 1899; 30 Stat. L., 1074. (An act making appropriations for sundry civil expenses of the government for the fiscal year ending June thirtieth, nineteen hundred, and for other purposes.)

That hereafter all standard, meander, township, and section lines of the public land surveys shall, as heretofore, be established under the direction and supervision of the Commissioner of the General Land Office, whether the lands to be surveyed are within or without reservations, except that where the exterior boundaries of public forest reservations are required to be coincident with standard, township or section lines such boundaries may, if not previously established in the ordinary course of the public land surveys, be established and marked under the supervision of the Director of the United States Geological Survey whenever necessary to complete the survey of such exterior boundaries. * * *

1902—Act of June 28, 1902; 32 Stat. L., 455. (An act making appropriations for sundry civil expenses of the government for the fiscal year ending June thirtieth, nineteen hundred and three, and for other purposes.)

Hereafter, in lieu of the specific estimates for personal services now required by law, there shall be submitted in the Annual Book of Estimates under each item of appropriation under "General Expenses of the Geological Survey," notes showing the number of persons employed and the rate of compensation paid to each from each of said appropriations during the fiscal year next preceding the fiscal year for which estimates are submitted.

Provided, That the purchase of professional and scientific books and periodicals needed for statistical purposes hereafter by the scientific divisions of the United States Geological Survey is hereby authorized to be made and paid for out of appropriations made for the said Survey.

1902—Act of May 16, 1902; 32 Stat. L., 741. (Joint resolution relating to publications of the Geological Survey.)

Resolved by the Senate and House of Representatives of the United States of America in Congress assembled, That hereafter the publications of the Geological Survey shall consist of the annual report of the Director, which shall be confined to one volume of royal octavo size; monographs, of quarto size; professional papers, of quarto size; bulletins, of ordinary octavo size; mineral resources, of ordinary octavo size; water-supply and irrigation papers, of ordinary octavo size; and such maps, folios, and atlases as may be required by existing law:

That hereafter the reports of the Geological Survey, except the annual report of the Director, shall be published in editions as recommended in each case by the Director and approved by the Secretary of the Interior, but not to exceed ten thousand copies.

That whenever the edition of any of the reports of the Survey shall have become exhausted, and the demand for it continues, there shall be published, on the requisition of the Secretary of the Interior, as many additional copies of the report as the Director of the Survey shall state will, in his judgment, be necessary to meet the demand.

That the bulletins and professional papers shall be distributed gratuitously, and not sold; and that of the number published one thousand copies shall be delivered to the Senate and two thousand copies shall be delivered to the House of Representatives for distribution.

That the Director of the Survey shall transmit to the Library of Congress two copies of every report of the Bureau as soon as the first delivery to the Survey is made, such copies to be additional to those received by the Library of Congress under existing law.

1903—Act of March 3, 1903; 32 Stat. L., 1146. (An Act making appropriations for sundry civil expenses of the government for the fiscal year ending June thirtieth, nineteen hundred and four, and for other purposes.)

And the Director of the Geological Survey shall hereafter distribute to public libraries that have not already received them such copies of sale publications as may remain on hand at the expiration of five years after date of delivery to the Survey document room, excepting a reserve number not to exceed two hundred copies.

1906—Act of June 30, 1906; 34 Stat. L., 727, 728. (An Act making appropriations for sundry civil expenses of the government for the fiscal year ending June thirtieth, nineteen hundred and seven, and for other purposes.)

The Secretary of the Interior is hereby authorized to permit scientific and other employees of the United States Geological Survey, employed in the field, to make assignments of their pay, under such regulations as he may prescribe, during such time as they may be in the employ of the United States Geological Survey. And the Secretary of the Interior is further authorized, in his discretion, under such regulations as he may prescribe, to reimburse the scientific and other employees for expenses incurred by them in the discharge of their duties in the field and paid from their personal funds.

The Director of the Geological Survey shall, if the regular map work of the Survey is in no wise interfered with thereby, hereafter furnish to any person, concern, institution, State or foreign government, that shall pay in advance the whole cost thereof with ten per centum added, transfers or copies of any cartographic or other engraved or lithographic data in the division of engraving and printing of the Survey, and the moneys received by the Director for such transfers or copies shall be deposited in the Treasury.

1909—Act of March 4, 1909; 35 Stat. L., 989. (An act making appropriations for sundry civil expenses of the government for the fiscal year ending June thirtieth, nineteen hundred and ten, and for other purposes.)

The Director of the Geological Survey shall hereafter furnish to any person, concern, or institution, in the interest of education and the dissemination of knowledge, that shall pay in advance the whole cost of material and services thereof, copies of any photographs or lantern slides in the possession of the United States Geological Survey; and the moneys received by the Director for the same shall be deposited in the United States Treasury.

1911—Act of March 1, 1911; 36 Stat. L., 961. (An Act to enable any state to coöperate with any other state or states, or with the United States, for the protection of the watersheds of navigable streams, and to appoint a commission for the acquisition of lands for the purpose of conserving the navigability of navigable rivers.)

Provided that before any lands are purchased by the National Forest Reservation Commission said lands shall be examined by the Geological Survey and a report made to the Secretary of Agriculture, showing that the control of such lands will promote or protect the navigation of streams on whose watersheds they lie.

1917—Act of June 12, 1917; 40 Stat. L., 144, 174. (An act making appropriations for sundry civil expenses of the government for the fiscal year ending June thirtieth, nineteen hundred and eighteen, and for other purposes.)

UNITED STATES GEOLOGICAL SURVEY

Office of Director: Director, $6,000; chief clerk, $2,500; chief disbursing clerk, $2,500; librarian, $2,000; photographer, $2,000; assistant photographers—one $900, one $720; clerk—one of class two, three of class one, one $1,000, four at $900 each; four copyists, at $720 each; watchmen—one $840, four at $720 each; janitor, $600; four messenger boys, at $480 each; in all, $35,340;

Scientific assistants: Geologists—two at $4,000 each, one $3,000, one $2,700; two paleontologists, at $2,000 each; chemist, $3,000; geographers—one $2,700, one $2,500; two topographers, at $2,000 each; in all, $29,900;

General expenses: For every expenditure requisite for and incident to the authorized work of the Geological Survey, including personal services in the District of Columbia and in the field, including the purchase, hire, maintenance, repair, and operation of motor-propelled and horse-drawn passenger-carrying vehicles for field use only by geologists, topographers, and engineers, to be expended under the regulations from time to time prescribed by the Secretary of the Interior, and under the following heads: *Provided,* That hereafter the purchase of supplies or the procurement of services outside the District of Columbia may be made in open market in the manner common among business men when the aggregate amount of the purchase does not exceed $50;

For pay of skilled laborers and various temporary employees, $20,000;

For topographic surveys in various portions of the United States, $350,000: *Provided,* That in expending this sum preference shall be given special topographic surveys of areas selected by the War Department and in securing such extra topographic data as are requested by the War Department in these or other areas;

For geologic surveys in the various portions of the United States, $350,000;

For continuation of the investigation of the mineral resources of Alaska, $100,000, to be immediately available;

For chemical and physical researches relating to the geology of the United States, including researches with a view of determining geological conditions favorable to the presence of deposits of potash salts, $40,000;

For preparation of the illustrations of the Geological Survey, $18,280;

For preparation of the report of the mineral resources of the United States, $75,000;

For gauging streams and determining the water supply of the United States, the investigation of underground currents and artesian wells, and the preparation of reports upon the best methods of utilizing the water resources, $175,000, of which $25,000 may be used to test the existence of artesian and other underground water supplies suitable for irrigation in the arid and semiarid regions by boring wells;

For purchase of necessary books for the library, including directories and professional and scientific periodicals needed for statistical purposes, $2,000;

For discovering, developing, protecting, and rendering more accessible, springs, streams, and water holes on arid public lands of the United States; for erecting and maintaining suitable and durable monuments and signboards; and for providing convenient and ready means, apparatus, and appliances by which water may be brought to the earth's surface, $10,000;

For engraving and printing geologic maps, $120,000;

For continuation of topographic surveys of the public lands that have been or may hereafter be designated as national forests, $75,000;

For the examination and classification of lands requisite to the determination of their suitability for enlarged homesteads, stock-raising homesteads, public watering places, and stock driveways, as required by the public land laws, to be immediately available, $150,000;

In all, United States Geological Survey, $1,550,520.

GOVERNMENT PRINTING OFFICE

For the United States Geological Survey: For engraving the illustrations necessary for the annual report of the director, and for the monographs, professional papers, bulletins, water-supply papers, and the report on mineral resources, and for printing and binding the same publications, of which sum not more than $45,000 may be used for engraving, $175,000.

APPENDIX 5

FINANCIAL STATEMENTS

Explanatory Note

Statements showing appropriations, receipts, expenditures and other financial data for a series of years constitute the most effective single means of exhibiting the growth and development of a service. Due to the fact that Congress has adopted no uniform plan of appropriations for the several services and that the latter employ no uniform plan in respect to the recording and reporting of their receipts and expenditures, it is impossible to present data of this character according to any standard scheme of presentation. In the case of some services the administrative reports contain tables showing financial conditions and operations of the service in considerable detail; in others financial data are almost wholly lacking. Careful study has in all cases been made of such data as are available, and the effort has been made to present the results in such a form as will exhibit the financial operations of the service in the most effective way that circumstances permit.

UNITED STATES GEOLOGICAL SURVEY.

APPROPRIATIONS: 1880 TO 1918, INCLUSIVE.

Appropriation	1880	1881	1882	1883	1884	1885	1886	1887
Expenses of the Geological Survey and the classification of public lands and the examination of the geologic structures, mineral resources, and products of the national domain	$100,000.00	$148,500.00	$148,500.00	$222,000.00	$240,000.00	$386,000.00	$400,000.00	$400,000.00
Salaries, Office of Geological Survey	6,000.00	6,000.00	6,000.00	34,940.00	34,940.00	35,340.00	35,540.00	35,540.00
Salaries, Scientific Assistants					64,700.00	67,700.00	67,700.00	67,700.00
Skilled Laborers and various Temporary Employees								
Topographic Surveys								
Geologic Surveys								
Paleontologic Researches								
Gaging Streams, Etc.								
Chemical and Physical Researches								
Mineral Resources of the United States								
Mineral Resources of Alaska								
Surveying National Forests								
Testing Fuel / Testing Structural Materials of U.S. / Investigating Mine Accidents / Inspecting Mines in Territories								
Geologic Maps of the United States								
Preparation of Illustrations								
Special Surveys and Miscellaneous Activities								
Books for the Library								
Printing and Binding								
Rent		1,500.00	1,500.00	1,500.00	1,500.00			10,000.00
Total	$106,000.00	$156,000.00	$156,000.00	$258,440.00	$341,140.00	$489,040.00	$503,240.00	$513,240.00

APPROPRIATIONS: 1880 TO 1918, INCLUSIVE—*Continued.*

Appropriation	1888	1889	1890	1891	1892	1893	1894	1895
Expenses of the Geological Survey and the classification of public lands and the examination of the geologic structures, mineral resources, and products of the national domain	$400,000.00							
Salaries, Office of Geological Survey	35,540.00	$35,540.00	$35,540.00	$35,540.00	$35,540.00	$35,540.00	$35,540.00	$31,634.57
Salaries, Scientific Assistants	67,700.00	67,700.00	67,700.00	67,700.00	67,700.00	33,332.88	29,900.00	29,900.00
Skilled Laborers and various Temporary Employees		15,000.00	15,000.00	15,000.00	15,000.00	10,000.00	13,000.00	13,000.00
Topographic Surveys		199,000.00	200,000.00	329,573.38	250,000.00	240,000.00	200,000.00	150,000.00
Geologic Surveys		100,000.00	100,000.00	115,000.00	115,000.00	50,000.00	70,000.00	100,000.00
Paleontologic Researches		40,000.00	40,000.00	40,000.00	40,000.00	10,000.00	10,000.00	10,000.00
Gaging Streams, Etc.								12,500.00
Chemical and Physical Researches		17,000.00	17,000.00	17,000.00	17,000.00	5,000.00	5,000.00	7,000.00
Mineral Resources of the United States		10,000.00	10,000.00	10,000.00	10,000.00	10,000.00	12,000.00	17,000.00
Mineral Resources of Alaska								
Surveying National Forests								
Testing Fuel; Testing Structural Materials of U.S.; Investigating Mine Accidents; Inspecting Mines in Territories								
Geologic Maps of the United States			45,000.00	70,000.00	60,000.00	20,000.00	55,000.00	65,000.00
Preparation of Illustrations		16,000.00	16,000.00	16,000.00	16,000.00	10,000.00	13,000.00	13,000.00
Special Surveys and Miscellaneous Activities		100,000.00	250,000.00	41,095.89				
Books for the Library (1)		5,000.00	5,000.00	5,000.00	2,500.00	2,000.00	2,000.00	3,000.00
Printing and Binding	122,000.00	122,000.00	68,000.00	68,000.00	63,000.00	48,000.00	35,000.00	35,000.00
Rent	10,000.00	10,000.00	10,000.00	13,200.00	13,200.00	14,200.00	14,200.00	14,200.00
Total	$635,240.00	$737,240.00	$879,240.00	$843,109.27	$704,940.00	$488,072.88	$494,640.00	$501,234.57

(1) From 1889 to 1904, inclusive, the amounts of appropriations appearing in this column include payments for transmission of public documents through the Smithsonian Exchange.

APPROPRIATIONS: 1880 TO 1918, INCLUSIVE—*Continued.*

Appropriation	1896	1897	1898	1899	1900	1901	1902	1903
Expenses of the Geological Survey and the classification of public lands and the examination of the geologic structures, mineral resources, and products of the national domain								
Salaries, Office of Geological Survey	$31,390.00	$31,390.00	$31,390.00	$31,390.00	$31,390.00	$31,390.00	$32,390.00	$32,390.00
Salaries, Scientific Assistants	29,900.00	29,900.00	29,900.00	29,900.00	29,900.00	29,900.00	29,900.00	29,900.00
Skilled Laborers and various Temporary Employees	13,000.00	13,000.00	13,000.00	13,000.00	13,000.00	13,000.00	16,000.00	20,000.00
Topographic Surveys	150,000.00	175,000.00	175,000.00	180,000.00	240,000.00	240,000.00	250,000.00	300,000.00
Geologic Surveys	100,000.00	100,000.00	100,000.00	110,000.00	110,000.00	150,000.00	150,000.00	150,000.00
Paleontologic Researches	10,000.00	10,000.00	10,000.00	10,000.00	10,000.00	10,000.00	10,000.00	10,000.00
Gaging Streams, Etc	24,500.00	50,000.00	50,000.00	50,000.00	70,000.00	100,000.00	100,000.00	200,000.00
Chemical and Physical Researches	8,500.00	7,000.00	7,000.00	7,000.00	7,000.00	10,000.00	20,000.00	20,000.00
Mineral Resources of the United States	18,000.00	20,000.00	20,000.00	20,000.00	30,000.00	50,000.00	50,000.00	50,000.00
Mineral Resources of Alaska (1)	5,000.00	5,000.00	5,000.00	5,000.00	25,000.00	60,000.00	60,000.00	60,000.00
Surveying National Forests (2)			150,000.00	150,000.00	130,000.00	130,000.00	130,000.00	13,0000.00
Testing Fuel; Testing Structural Materials of U.S.; Investigating Mine Accidents; Inspecting Mines in Territories								
Geologic Maps of the United States	65,000.00	60,000.00	60,000.00	60,000.00	62,500.00	80,000.00	75,000.00	100,000.00
Preparation of Illustrations	13,000.00	13,000.00	13,000.00	14,000.00	14,000.00	14,000.00	16,280.00	18,280.00
Special Surveys and Miscellaneous Activities	200,000.00	208,000.00	279,150.00	55,889.60	4,338.45	12,000.00	7,830.00	7,200.00
Books for the Library (3)	4,230.60	2,000.00	4,319.25	4,997.75	6,912.44	7,620.00	5,000.00	6,000.00
Printing and Binding	41,000.00	37,000.00	37,000.00	77,000.00	37,000.00	37,000.00	45,000.00	215,000.00
Rent	14,600.00	15,400.00	15,400.00	17,400.00	17,400.00	22,400.00	25,400.00	28,400.00
Total	$728,120.60	$776,690.00	$1,000,159.25	$835,577.35	$838,440.89	$997,310.00	$1,022,800.00	$1,377,170.00

(1) Appropriations for the years 1896 to 1901, inclusive, were for "Coal and gold resources of Alaska."

(2) Appropriations for the years 1898 to 1910, inclusive, were for "Surveying Forest Reserves."

(3) From 1889 to 1904, inclusive, the amounts of appropriations appearing in this column include payments for transmission of public documents through the Smithsonian Exchange.

APPROPRIATIONS: 1880 TO 1918, INCLUSIVE—*Continued.*

Appropriation	1904	1905	1906	1907	1908	1909	1910	1911
Expenses of the Geological Survey and the classification of public lands and the examination of the geologic structures, mineral resources, and products of the national domain	....	....	....	....	....	....	....	....
Salaries, Office of Geological Survey	$32,740.00	$32,740.00	$32,740.00	$35,340.00	$35,340.00	$35,340.00	$34,860.00	$34,860.00
Salaries, Scientific Assistants	29,900.00	29,900.00	29,900.00	29,900.00	29,900.00	29,900.00	29,900.00	29,900.00
Skilled Laborers and various Temporary Employees	20,000.00	20,000.00	20,000.00	20,000.00	20,000.00	20,000.00	20,000.00	20,000.00
Topographic Surveys	300,000.00	300,000.00	350,000.00	350,000.00	300,000.00	300,000.00	350,000.00	350,000.00
Geologic Surveys	150,000.00	175,000.00	200,000.00	200,000.00	200,000.00	200,000.00	225,000.00	300,000.00
Paleontologic Researches (1)	10,000.00	10,000.00	10,000.00	10,000.00	10,000.00	10,000.00	....	....
Gaging Streams, Etc.	200,000.00	200,000.00	200,000.00	150,000.00	100,000.00	100,000.00	100,000.00	150,000.00
Chemical and Physical Researches (2)	20,000.00	20,000.00	20,000.00	20,000.00	20,000.00	20,000.00	20,000.00	20,000.00
Mineral Resources of the United States	60,000.00	50,000.00	100,000.00	75,000.00	75,000.00	75,000.00	75,000.00	75,000.00
Mineral Resources of Alaska	60,000.00	80,000.00	80,000.00	80,000.00	80,000.00	80,000.00	90,000.00	90,000.00
Surveying National Forests (3)	130,000.00	130,000.00	130,000.00	100,000.00	100,000.00	75,000.00	75,000.00	75,000.00
Testing Fuel (4)	....	60,000.00	227,000.00	250,000.00	250,000.00	250,000.00	100,000.00	....
Testing Structural Materials of U.S. (4)	....	....	12,500.00	100,000.00	100,000.00	100,000.00	100,000.00	....
Investigating Mine Accidents (4)	....	....	....	....	....	150,000.00	150,000.00	....
Inspecting Mines in Territories (4)	....	....	....	....	....	7,350.00	7,350.00	....
Geologic Maps of the United States	100,000.00	100,000.00	100,000.00	100,000.00	100,000.00	100,000.00	100,000.00	100,000.00
Preparation of Illustrations	18,280.00	18,280.00	18,280.00	18,280.00	18,280.00	18,280.00	18,280.00	18,280.00
Special Surveys and Miscellaneous Activities	10,857.77	10,500.00	7,062.30	....	....	17,810.00	425.00	....
Books for the Library	6,000.00	2,000.00	2,000.00	2,000.00	2,000.00	2,000.00	2,000.00	2,000.00
Printing and Binding	215,000.00	215,000.00	215,000.00	185,000.00	185,000.00	175,000.00	175,000.00	175,000.00
Rent	28,400.00	30,400.00	30,400.00	31,900.00	34,900.00	34,900.00	37,400.00	37,400.00
Total	$1,391,177.77	$1,483,820.00	$1,784,882.30	$1,757,420.00	$1,660,420.00	$1,800,580.00	$1,710,215.00	$1,477,440.00

(1) Combined with the appropriations for the geological survey beginning with the year 1910.

(2) Includes researches relating to deposits of potash salts during the last seven years.

(3) Appropriations for the years 1898 to 1910, inclusive, were for "Surveying Forest Reserves."

(4) Bureau of Mines was created in 1910, and thereafter appropriations for these activities were made under that unit.

APPROPRIATIONS: 1880 TO 1918, INCLUSIVE—*Continued.*

Appropriation	1912	1913	1914	1915	1916	1917	1918	Total
Expenses of the Geological Survey and the classification of public lands and the examination of the geologic structures, mineral resources, and products of the national domain....	....	....	....	....	....	....	....	$2,445,000.00
Salaries, Office of Geological Survey..	$35,340.00	$35,340.00	$35,340.00	$35,340.00	$35,340.00	$35,340.00	$35,340.00	1,249,174.57
Salaries, Scientific Assistants.........	29,900.00	29,900.00	29,900.00	29,900.00	29,900.00	29,900.00	29,900.00	1,387,132.88
Skilled Laborers and various Temporary Employees...............	20,000.00	20,000.00	20,000.00	20,000.00	20,000.00	20,000.00	20,000.00	510,000.00
Topographic Surveys...............	350,000.00	350,000.00	350,000.00	350,000.00	350,000.00	350,000.00	350,000.00	8,328,573.38
Geologic Surveys.................	300,000.00	300,000.00	300,000.00	400,000.00	350,000.00	350,000.00	350,000.00	5,620,000.00
Paleontologic Researches..........	....	....	....	....	....	....	....	330,000.00
Gaging Streams, Etc...............	150,000.00	150,000.00	150,000.00	150,000.00	150,000.00	150,000.00	175,000.00	2,932,000.00
Chemical and Physical Researches (1).	40,000.00	40,000.00	40,000.00	40,000.00	40,000.00	40,000.00	40,000.00	611,500.00
Mineral Resources of the United States.	75,000.00	75,000.00	75,000.00	75,000.00	75,000.00	75,000.00	75,000.00	1,447,000.00
Mineral Resources of Alaska.........	100,000.00	90,000.00	100,000.00	100,000.00	100,000.00	100,000.00	100,000.00	1,555,000.00
Surveying National Forests..........	75,000.00	75,000.00	75,000.00	75,000.00	75,000.00	75,000.00	75,000.00	2,160,000.00
Testing Fuel................ (2)	....	....	....	....	....	....	....	1,137,000.00
Testing Structural Materials of U.S. (2)	....	....	....	....	....	....	....	412,500.00
Investigating Mine Accidents.... (2)	....	....	....	....	....	....	....	300,000.00
Inspecting Mines in Territories.. (2)	....	....	....	....	....	....	....	14,700.00
Geologic Maps of the United States...	110,000.00	110,000.00	110,000.00	110,000.00	110,000.00	110,000.00	120,000.00	2,457,500.00
Preparation of Illustrations.........	18,280.00	18,280.00	18,280.00	18,280.00	18,280.00	18,280.00	18,280.00	489,760.00
Special Surveys and Miscellaneous Activities...................	....	....	....	....	....	....	....	1,372,159.01
Books for the Library..............	2,000.00	2,000.00	2,000.00	2,000.00	2,000.00	2,000.00	160,000.00	99,580.04
Printing and Binding...............	145,000.00	165,000.00	175,000.00	175,000.00	175,000.00	175,000.00	2,000.00	3,812,000.00
Rent..........................	37,400.00	37,400.00	37,400.00	40,000.00	40,000.00	40,000.00	175,000.00	769,300.00
Total........................	$1,487,920.00	$1,497,920.00	$1,517,920.00	$1,620,520.00	$1,570,520.00	$1,570,520.00	$1,725,520.00	$39,439,879,88

(1) Includes researches relating to deposits of potash salts during the last seven years.

(2) Bureau of Mines was created in 1910, and thereafter appropriations for these activities were made under that unit.

STATE CONTRIBUTIONS: COÖPERATIVE TOPOGRAPHIC AND HYDROGRAPHIC WORK.[1]

States	Prior to June 30, 1909[2]	1909-10	1910-11	1911-12	1912-13	1913-14	1914-15	1915-16	Total
Alabama:									
Topographic Work....	$6,000.00	....	$800.00	....	....	....	$3,032.78	....	$9,832.78
Hydrographic Work...	994.00	....	....	....	....	....	170.00	$200.00	1,364.00
Arizona:									
Hydrographic Work...	....	....	....	....	$3,000.00	$3,000.00	3,900.00	3,960.00	13,860.00
California:									
Topographic Work....	80,000.00	$14,000.00	14,000.00	$21,000.00	17,600.00	16,500.00	14,000.00	14,000.00	191,100.00
Hydrographic Work...	79,486.73	9,000.00	9,000.00	25,000.00	28,586.00	16,500.00	13,200.00	16,560.00	197,332.73
Colorado:									
Hydrographic Work...	1,487.97	....	545.00	167.10	....	....	480.00	485.00	3,165.07
Connecticut:									
Topographic Work....	24,599.00	....	....	....	....	....	....	....	24,599.00
Hydrographic Work...	....	....	....	....	1,000.00	1,000.00	1,000.00	1,555.00	4,555.00
Idaho:									
Topographic Work....	....	....	....	....	700.00	....	....	....	700.00
Hydrographic Work...	2,000.00	2,000.00	5,000.00	12,000.00	5,000.00	9,000.00	1,900.00	....	36,900.00
Illinois:									
Topographic Work....	36,000.00	17,500.00	13,750.00	13,750.00	8,000.00	11,500.00	9,000.00	12,000.00	121,500.00
Hydrographic Work...	8,675.94	2,000.00	3,000.00	2,191.37	....	....	3,400.00	2,355.00	21,622.31
Iowa:									
Topographic Work....	3,500.00	1,750.00	1,750.00	2,350.00	2,650.00	1,750.00	1,950.00	1,750.00	17,450.00
Hydrographic Work...	....	....	....	....	....	500.00	450.00	500.00	1,450.00
Kansas:									
Hydrographic Work...	413.03	....	....	....	....	....	....	....	413.03
Kentucky:									
Topographic Work....	37,500.00	5,000.00	10,000.00	10,000.00	10,000.00	10,000.00	10,000.00	10,000.00	102,500.00
Hydrographic Work...	....	....	....	....	....	....	670.00	315.00	985.00
Louisiana:									
Topographic Work....	4,000.00	20,000.00	....	....	....	....	....	....	24,000.00
Maine:									
Topographic Work....	26,700.00	3,500.00	4,500.00	4,700.00	3,726.42	4,900.00	4,000.00	5,000.00	57,026.42
Hydrographic Work...	37,505.07	3,300.00	1,000.00	1,350.00	....	....	75.00	4,515.00	47,745.07
Maryland:									
Topographic Work....	46,550.00	3,000.00	2,000.00	1,825.00	....	....	....	....	53,375.00
Hydrographic Work...	3,400.00	....	....	....	....	....	....	....	3,400.00
Massachusetts:									
Topographic Work....	40,000.00	....	....	....	....	....	....	....	40,000.00
Hydrographic Work...	1,050.00	1,050.00	825.00	1,000.00	3,000.00	3,000.00	3,350.00	2,520.00	15,795.00
Michigan:									
Topographic Work....	13,700.00	2,000.00	2,000.00	2,000.00	2,000.00	2,500.00	2,500.00	15,000.00	41,700.00

[1] In addition to the amounts appearing in this statement, funds have been appropriated by states for coöperative geologic surveys, complete records of which are not available.

[2] Amounts in this column are taken from pamphlet entitled "Coöperation between the United States and Various States in Topographic, Hydrographic and Geologic Work," published by the Survey in 1910.

State Contributions: Coöperative Topographic and Hydrographic Work[1]—*Continued.*

States	Prior to June 30, 1909[2]	1909-10	1910-11	1911-12	1912-13	1913-14	1914-15	1915-16	Total
Minnesota:									
Topographic Work	....	$10,000.00	$8,000.00	$10,000.00	$10,362.48	$10,000.00	$4,500.00	$4,250.00	$57,112.48
Hydrographic Work	$18,400.00	17,500.00	....	16,000.00	3,650.00	3,500.00	2,100.00	2,400.00	63,550.00
Mississippi:									
Topographic Work	31,400.00	....	....	....	....	....	....	....	31,400.00
Missouri:									
Topographic Work	6,000.00	4,500.00	4,000.00	6,000.00	2,500.00	10,000.00	5,221.74	4,000.00	42,221.74
Montana:									
Hydrographic Work	1,405.00	....	....	3,000.00	2,500.00	....	2,500.00	2,875.00	12,280.00
Nebraska:									
Topographic Work	....	....	....	1,500.00	2,000.00	2,000.00	....	1,750.00	7,250.00
Hydrographic Work	1,700.00	....	....	....	....	....	....	....	1,700.00
Nevada:									
Hydrographic Work	1,250.00	....	....	....	385.00	2,500.00	1,800.00	3,140.00	9,075.00
New Hampshire:									
Hydrographic Work	1,300.00	....	....	....	....	....	....	....	1,300.00
New Jersey:									
Topographic Work	19,670.00	....	....	....	....	....	....	....	19,670.00
New Mexico:									
Hydrographic Work	2,500.00	2,500.00	3,170.93	4,258.15	14,174.00	12,000.00	7,200.00	....	45,803.08
New York:									
Topographic Work	233,600.00	7,000.00	10,000.00	10,000.00	10,000.00	10,000.00	10,000.00	10,000.00	300,600.00
Hydrographic Work	21,373.91	5,000.00	11,089.69	13,476.30	11,500.00	11,500.00	11,900.00	13,800.00	99,639.90
North Carolina:									
Topographic Work	29,200.00	2,500.00	....	....	....	....	....	....	31,700.00
Hydrographic Work	576.25	....	....	....	....	....	....	....	576.25
North Dakota:									
Hydrographic Work	....	....	....	....	....	....	350.00	500.00	850.00
Ohio:									
Topographic Work	183,200.00	15,000.00	25,000.00	28,400.00	25,000.00	25,000.00	35,000.00	23,900.00	360,500.00
Hydrographic Work	1,800.00	....	....	....	....	....	....	....	1,800.00
Oklahoma:									
Topographic Work	14,000.00	....	....	1,000.00	1,000.00	1,000.00	1,000.00	....	18,000.00
Oregon:									
Topographic Work	10,000.00	2,500.00	2,500.00	15,000.00	15,000.00	18,000.00	18,000.00	4,000.00	85,000.00
Hydrographic Work	15,000.00	2,500.00	2,500.00	13,419.47	7,470.00	10,000.00	11,000.00	7,300.00	69,189.47
Pennsylvania:									
Topographic Work	147,500.00	4,238.00	5,183.00	5,328.00	5,528.19	6,816.55	4,000.00	....	178,593.74
Hydrographic Work	2,188.35	....	....	....	....	....	....	....	2,188.35

[1] In addition to the amounts appearing in this statement, funds have been appropriated by states for coöperative geologic surveys, complete records of which are not available.

[2] Amounts in this column are taken from pamphlet entitled "Coöperation between the United States and Various States in Topographic, Hydrographic and Geologic Work," published by the Survey in 1910.

STATE CONTRIBUTIONS: COÖPERATIVE TOPOGRAPHIC AND HYDROGRAPHIC WORK[1]—*Continued.*

States	Prior to June 30, 1909[2]	1909-10	1910-11	1911-12	1912-13	1913-14	1914-15	1915-16	Total
Rhode Island:									
Topographic Work....	$5,000.00	....	....	....	....	....	....	....	$5,000.00
Hydrographic Work...	1,800.00	....	....	....	....	....	....	....	1,800.00
South Dakota:									
Hydrographic Work...	....	....	....	....	....	....	$960.00	....	960.00
Tennessee:									
Topographic Work....	....	....	$4,250.00	....	....	....	....	....	4,250.00
Hydrographic Work...	....	....	....	$718.22	....	....	....	....	718.22
Texas:									
Topographic Work....	5,000.00	$20,000.00	....	....	....	....	35,000.00	$35,000.00	95,000.00
Hydrographic Work...	....	....	....	....	....	....	....	8,500.00	8,500.00
Utah:									
Hydrographic Work...	2,950.00	2,000.00	2,000.00	2,521.47	$2,000.00	$5,500.00	4,100.00	5,015.00	26,086.47
Vermont:									
Topographic Work....	....	....	....	....	....	2,000.00	2,000.00	2,500.00	6,500.00
Hydrographic Work...	....	1,000.00	1,000.00	1,000.00	1,000.00	1,200.00	900.00	1,440.00	7,540.00
Virginia:									
Topographic Work....	1,750.00	1,750.00	1,750.00	4,250.00	4,250.00	4,250.00	4,500.00	4,500.00	27,000.00
Hydrographic Work...	987.56	....	....	....	....	....	....	....	987.56
Washington:									
Topographic Work....	....	10,000.00	12,500.00	13,750.00	13,750.00	12,500.00	11,500.00	11,500.00	85,500.00
Hydrographic Work...	6,150.00	5,000.00	5,000.00	4,419.31	4,200.00	5,000.00	5,950.00	6,020.00	41,739.31
West Virginia:									
Topographic Work....	117,000.00	15,000.00	12,000.00	12,000.00	12,000.00	14,331.41	6,758.20	7,639.39	196,729.00
Hydrographic Work...	....	....	....	....	132.00	....	400.00	....	532.00
Wisconsin:									
Topographic Work....	....	....	....	....	....	....	4,000.00	....	4,000.00
Hydrographic Work...	3,361.00	....	....	....	....	8,500.00	8,800.00	6,000.00	26,661.00
Wyoming:									
Hydrographic Work...	....	....	....	200.00	265.00	....	1,600.00	5,010.00	7,075.00
Hawaii:									
Topographic Work....	....	....	15,000.00	15,000.00	15,200.00	16,000.00	837.35	....	62,037.35
Hydrographic Work...	....	....	5,000.00	12,000.00	20,000.00	15,000.00	15,400.00	20,000.00	87,490.00
Total for both Topographic and Hydrographic Work.........	$1,339,623.81	$212,088.00	$198,113.62	$290,574.39	$269,129.09	$286,747.96	$290,355.07	$281,844.39	$3,168,476.33
Total: Topographic Work....	$1,121,869.00	$159,238.00	$148,983.00	$177,853.00	$161,267.09	$179,047.96	$186,800.07	$166,789.39	$2,301,847.51
Total: Hydrographic Work...	$217,754.81	$52,850.00	$49,130.62	$112,721.39	$107,862.00	$107,700.00	$103,555.00	$115,055.00	$866,628.82

[1] In addition to the amounts appearing in this statement, funds have been appropriated by states for coöperative geologic surveys, complete records of which are not available.

[2] Amounts in this column are taken from pamphlet entitled "Coöperation between the United States and Various States in Topographic, Hydrographic and Geologic Work," published by the Survey in 1910.

APPENDIX 6

BIBLIOGRAPHY

Explanatory Note

The bibliographies appended to the several monographs aim to list only those works which deal directly with the services to which they relate, their history, activities, organization, methods of business, problems, etc. They are intended primarily to meet the needs of those persons who desire to make a further study of the services from an administrative standpoint. They thus do not include the titles of publications of the services themselves, except in so far as they treat of the services, their work and problems. Nor do they include books or articles dealing merely with technical features, other than administrative, of the work of the services. In a few cases explanatory notes have been appended where it was thought they would aid in making known the character or value of the publication to which they relate.

After the completion of the series the bibliographies may be assembled and separately published as a Bibliography of the Administrative Branch of the National Government.

BIBLIOGRAPHY

UNITED STATES GEOLOGICAL SURVEY

CONTENTS

BIBLIOGRAPHY

UNITED STATES GEOLOGICAL SURVEY

Bibliographies

Note: Dashes at the beginning of an item indicate repetition of words in the preceding author entry. Thus, in item 6, the dashes indicate U. S. Geological survey.

1 Prime, Frederick. Catalogue of official reports upon geological surveys of the United States and territories and of British North America. (*In* American institute of mining engineers. Transactions. May, 1878 to Feb. 1879. Easton, Pa., 1879. v. 7, 1879. p. 455-525)

[Classified by states and provinces]

2 Schmeckebier, Laurence F. Catalogue and index of the publications of the Hayden, King, Powell, and Wheeler surveys: namely, Geological and geographical survey of the territories, Geological exploration of the fortieth parallel, Geographical and geological surveys of the Rocky Mountain region, Geographical surveys west of the one hundredth meridian. Washington, Govt. print. off., 1904. 208 p. (U. S. Geological survey. Bulletin no. 222) Issued also as House doc. no. 606, 58th Cong., 2d sess.

3 U. S. *Geographical surveys west of the 100th meridian.* List of reports and maps of the United States Geographical surveys west of the 100th meridian. 2d ed. Washington, Govt. print. off., 1881. 74 p.

4 —— *Geological and geographical survey of the territories.* Catalogue of the publications of the U. S.

Geological and geographical survey of the territories. F. V. Hayden, United States geologist. 3d ed., rev. to December 31, 1878. Washington, Govt. print. off., 1879. 54 p.

5 U. S. *Geological survey.* The publications of the United States Geological survey (not including topographic maps) December, 1916. Washington, Govt. print. off., 1916. 168 p.

6 —— —— Technologic publications of the United States Geological survey. [Washington, 1910] 3 p.

7 —— —— Topographic maps and folios and geologic folios published by the United States Geological survey. Washington, U. S. Geological survey, 1914. 131 p.

8 —— *Superintendent of documents.* Publications of the United States Geological survey, for sale by the superintendent of documents, Washington, D. C. [9th ed.] [Washington, Govt. print. off.] 1916. 60 p. (Price list 15—9th ed.)

9 —— —— Reports of explorations printed in the documents of the United States government. [A contribution toward a bibliography] Comp. by Adelaide R. Hasse. Washington, Govt. print. off., 1899. 90 p.

[Includes references concerning the early surveys which preceded the Geological survey]

10 Warman, Philip C. Bibliography and index of the publications of the United States Geological survey with the laws governing their printing and distribution. Washington, Govt. print. off., 1893. 495 p. (U. S. Geological survey. Bulletin no. 100)

11 —— Catalogue and index of the publications of the United States Geological survey, 1880 to 1901. Washington, Govt. print. off., 1901. 858 p. (U. S. Geological survey. Bulletin no. 177)

12 —— Catalogue and index of the publications of the United States Geological survey 1901 to 1903. Wash-

ington, Govt. print. off., 1903. 234 p. (U. S. Geological survey. Bulletin no. 215)
Supplemental to the "Catalogue and index" pub. in 1901, as Bulletin no. 177.

OFFICIAL PUBLICATIONS RELATING TO EARLY SURVEYS

13 Featherstonhaugh, George W. Geological report of an examination made in 1834 of the elevated country between the Missouri and Red rivers. Washington, Printed by Gales and Seaton, 1835. 97 p. (23d Cong., 2d sess. House. Doc. 151) Serial no. 269. Report addressed to Lieut. Col. J. J. Abert, U. S. topographical engineers.

14 Merrill, George P. The era of national surveys . . . 1870-1879. (*In* U. S. National museum. Report, 1904. Washington, 1906. p. 551-624)

[A chapter from "Contributions to the history of American geology" by George P. Merrill. Contains portraits of the noted geologists of that decade. Excellent account of the various geological and geographical surveys which were afterward consolidated under the U. S. Geological survey]

15 U. S. *Congress. House. Committee on public lands.* Geographical and geological surveys west of the Mississippi. Report . . . May 26, 1874. Washington, Govt. print. off., 1874. 91 p. (43d Cong., 1st sess. House. Rept. no. 612) Serial no. 1626

16 —— —— —— Public surveys. Testimony . . . May 11, 1878. Washington, Govt. print. off., 1878. 46 p. (45th Cong., 2d sess. House. Mis. doc. no. 55) Serial no. 1818

[Major Powell, Professor Hayden and Lieut. Wheeler were among the witnesses]

17 —— *Dept. of the interior.* Cost of geographical surveys. Letter from [J. W. Powell] relating to the cost of geographical surveys. [Washington, Govt. print. off., 1879] 6 p. (45th Cong., 3d sess. House. Ex. doc. no. 72) Serial no. 1858

18 U. S. *Dept. of the interior.* Geological and geographical surveys . . . report of Professor Hayden upon geological and geographical surveys. April 19, 1878. [Washington Govt. print. off., 1878] 22 p. fold. map. (45th Cong., 2d sess. House. Ex. doc. no. 81)

Serial no. 1809

19 —— *Geographical and geological survey of the Rocky Mountain region.* Geological and geographical surveys. A report of Professor Powell in regard to surveys, in response to a resolution of the House of Representatives . . . Washington, Govt. print. off., 1878. 19 p. fold. map. (45th Cong., 2d sess. House. Ex. Doc. no. 80) Serial no. 1809

20 —— —— Report on the methods of surveying the public domain . . . at the request of the National academy of sciences. By J. W. Powell, 1878. Washington, Govt. print. off., 1878. 16 p.

21 —— —— Report on the Geographical and geological survey of the Rocky Mountain region, by J. W. Powell. Washington, Govt. print. off., 1877. 19 p.

22 —— *Geographical surveys west of the 100th meridian.* Annual report . . . [1873-1884] by George M. Wheeler, captain corps of engineers, U. S. A. Washington, Govt. print. off., 1873-1884. 12 v.
Reports are from Appendices to the Annual reports of the U. S. Chief of engineers for 1873-1884.

23 —— —— Report upon United States Geographical surveys west of the one hundredth meridian, in charge of First Lieut. Geo. M. Wheeler . . . under the direction of the Chief of engineers, U. S. army. Washington, Govt. print. off., 1875-1889. 7 v. in 8.

24 U. S. *Geological and geographical survey of the territories.* [1st]-12th annual report of the United States Geological and geographical survey of the territories . . . By F. V. Hayden, United States geolo-

gist . . . Washington, Govt. print. off., 1867-1883. 12 v. in 14.

Atlas accompanies v. 12. No more published. Survey was discontinued by act of Congress of March 3, 1879, and the U. S. Geological survey established in its stead.

25 U. S. *Geological and geographical survey of the territories.* Sketch of the origin and progress of the United States Geological and geographical survey of the territories. F. V. Hayden, U. S. geologist-in-charge. Washington, D. C., Darby & Duvall, printers, 1877. 15 p.

26 —— *Geological exploration of the fortieth parallel.* Report . . . made by order of the secretary of war . . . under the direction of . . . A. A. Humphreys, chief of engineers, by Clarence King . . . Washington, Govt. print. off., 1870-1880. 7 v. and atlas.

27 —— *Treasury dept.* Cost of certain surveys. Letter . . . transmitting a communication from the . . . Coast and geodetic survey relative to the cost of certain classes of work of the Survey. Washington, Govt. print. off., 1879. 4 p. (45th Cong., 3d sess. House. Ex. doc. no. 29) Serial no. 1852

28 —— *War dept.* Duplication of surveys of the Mississippi river. Letter . . . concerning duplication of the surveys of the Mississippi river by the Coast survey. Washington, Govt. print. off., 1879. 3 p. (45th Cong., 3d sess. House. Ex. doc. no. 19) Serial no. 1852

29 —— —— Geographical and geological surveys west of the Mississippi . . . report . . . May 2, 1874. [Washington, Govt. print. off., 1874] 15 p. (43d Cong., 1st sess. House. Ex. doc. no. 240)

Serial no. 1614

30 —— —— Letter communicating information in relation to the surveys of the territories . . . and transmitting the views of the Chief of engineers concerning a

report of the National academy of sciences. Jan. 13, 1879. [Washington, Govt. print. off., 1879] 17, 5 p. (45th Cong., 3d sess. Senate. Ex. doc. no. 21) Serial no. 1828

31 U. S. *War dept.* Lieutenant Wheeler's geographical surveys. Letter . . . relative to the continuance of the geographical explorations and surveys west of the one hundredth meridian. [Washington, Govt. print. off., 1875] 3 p. (43d Cong., 2d sess. House. Ex. doc. no. 109) Serial no. 1648

32 —— —— Reports of explorations and surveys to ascertain the most practicable and economical route for a railroad from the Mississippi River to the Pacific Ocean, made under the direction of the secretary of war in 1853-[6] Washington, G. O. P. Nicholson, printer, 1855-1860. 12 v. in 13. plates, maps. [Vols. 1:11: 33d Cong., 2d sess. House. Ex. doc. no. 91. Vol. 12, pts. 1-2: 36th Cong., 1st sess. House. Ex. doc. no. 56. Serial no. 791-801, 1054-1055.
The reports of the Pacific railroad surveys were prepared under supervision of the Engineer department]

33 —— —— Surveys by the War department. Letter . . . giving information concerning the surveys conducted by the department in the last ten years. May 10, 1878. [Washington, Govt. print. off., 1878] 8 p. fold. map. (45th Cong., 2d sess. House. Ex. doc. no. 88) Serial no. 1809
I. Geological exploration of the 40th parallel.
II. Geographical surveys west of the 100th meridian.

Official Publications Relating to the Geological Survey

34 Documents relating to the organization of the U. S. Geological survey, 1874-1880. 1 v.

[Brought together and bound in the library of the Geological survey, with typewritten title-page and table of contents]

35 Herbert, Hilary H. Restricting the work and publications of the Geological survey, and for other purposes. Report [to accompany bill H. R. 8320] [Washington] 1886. 16 p. ([U. S.] 49th Cong., 1st sess. House. Rept. no. 2214) Serial no. 2441

[A criticism of extravagance of service in amount provided, in plan of work, etc. Recommends following English plan which leaves discussion and research to private enterprise. Includes summary table showing annual appropriations for geological works of certain foreign countries]

36 India. *Survey of India dept.* Report on the United States government surveys by Maj. Genl. J. Waterhouse . . . Calcutta, Office of the superintendent, government printing, 1906. 47 p. xiv pl.

[Includes an interesting account of the Geological survey]

37 Powell, John W. On the organization of scientific work of the general government . . . Extracts from the testimony taken by the Joint Commission of the Senate and House to consider the present organizations of the Signal service, Geological survey . . . Washington, Govt. print. off., 1885. [177] p.

[Defense of the Geological survey; reply to criticisms made by Alexander Agassiz and others concerning the bulk and relative value of Survey publications. Short summary of methods of conducting government surveys . . . in various countries]

38 U. S. *Commission to consider the present organizations of the Signal service, Geological survey, Coast and geodetic survey, and the Hydrographic office, etc.* Report . . . June 8, 1886. Washington, Govt. print. off., 1886. 125 p. (49th Cong., 1st sess. Senate. Rept. no. 1285) Serial no. 2361

39 —— —— Testimony before the Joint Commission . . . Washington, Govt. print. off., 1886. 37, 1104 p. (49th Cong., 1st sess. Senate. Mis. doc. no. 82) Serial no. 2345

[Testimony of Major Powell, Director of the Survey, scattered through the volume]

40 U. S. *Congress. House. Committee on agriculture.* Transfer of Fish commission and Geological survey to Department of agriculture. Report [to accompany H. R. 6233] May 9, 1894. [Washington, Govt. print. off., 1894] 5 p. (53d Cong., 2d sess. House Rept. no. 863) Serial no. 3271

[Report favors transfer of Geological survey to the Department of Agriculture under the name "Bureau of Geological survey"]

41 —— —— *Committee on expenditures in Interior dept.* Topographic and hydrographic surveys. No. 1. Hearings . . . on H. R. 13457, a bill to provide for a more equitable means of distribution of . . . work of the United States Geological survey among the several states . . . Washington, Govt. print. off., 1914. 30 p.

[Preceded by a letter from the director of the Geological survey describing the manner in which the system recommended in the bill would work out, and stating objections to it]

42 —— —— *Committee on interstate and foreign commerce.* Report . . . favoring H. 9056, to continue investigation of water resources [by Geological survey] Feb. 13, 1912. Washington, Govt. print. off., 1912. 3 p. (62d Cong., 2d sess. House. Rept. 319) Serial no. 6129

43 —— —— *Committee on irrigation of arid lands.* Report . . . favoring House bill 6246, to provide topographic surveys and to mark elevations on arid lands. Feb. 18, 1896. [Washington, Govt. print. off., 1896] 6 p. (54th Cong., 1st sess. House. Rept. 424) Serial no. 3458

44 —— —— *Committee on mines and mining.* Investigation of water resources in the United States . . . Report. [To accompany H. R. 23583] [Washington, Govt. print. off., 1907] 8 p. (59th Cong., 2d sess. House. Rept. 6408) Serial no. 5064

[Favorable report on bill which provides authority to the Geological survey for coöperation with states in water resources work]

45 U. S. *Congress. House. Committee on printing.* Report . . . amending H. J. R. 259, for change of method of distribution of bulletins and professional papers issued by Geological survey . . . Feb. 26, 1903. [Washington, Govt. print. off., 1903] 2 p. (57th Cong., 2d sess. House. Rept. 3912)
Serial no. 4415

46 —— —— *Committee on public buildings and grounds.* Report . . . favoring H. 28377, to authorize erection of public building for Survey and other offices. Feb. 23, 1909. Washington, Govt. print. off., 1909. 1 p. (60th Cong., 2d sess. House. Rept. 2284)
Serial no. 5384

47 —— —— *Senate. Committee on the Geological survey.* Investigation of the water resources of the United States . . . Report. [To accompany S. 7359] [Washington, Govt. print. off., 1907] 18 p. (59th Cong., 2d sess. Senate. Rept. 6667) Serial no. 5060

[Favorable report on bill authorizing coöperation of Geological survey with various states in hydrographic and other surveys, under supervision of the Director of the survey]

48 —— —— *Select committee on methods of business and work in the executive departments.* The U. S. Geological survey. (*In its* Report. Washington, 1888. pt. 2, p. 377-490) Serial no. 2521

[Report on the organization, business methods, and work of the Survey; accompanied by exhibits. Published as part of Senate Report 507, 50th Cong., 1st sess. This report was known as the Cockrell committee report]

49 U. S. *Dept. of the interior.* Estimate of appropriation for building for Survey. Jan. 27, 1909. Washington, Govt. print. off., 1909. 5 p. (60th Cong., 2d sess. House. Doc. 1372) Serial no. 5559

50 U. S. *Joint commission on laws organizing executive departments.* Geological survey. (*In its* Organiza-

tion of the executive departments . . . Report. [Washington, 1893] p. 140-4)

Published as part of House report no. 88, or Senate report no. 47 of the 53d Congress, 1st session. Known as the Dockery-Cockrell commission report. Serial no. 3158 and 3148

51 U. S. *Joint commission on laws organizing executive departments.* Geological survey. (*In its* References to laws organizing executive departments . . . [Washington, 1893] p. 103-6) Serial no. 3158 and 3148

[Part of House report no. 49 or Senate report no. 41 of the 53d Congress, 1st session. Known as the Dockery-Cockrell commission report. Contains extracts from laws creating and appropriating for the various offices in the Survey]

52 —— —— Report . . . [recommending reference of H. R. 8504, to improve public surveys, to Committee on public lands] Feb. 28, 1895. [Washington, Govt. print. off., 1895] 29 p. (53d Cong., 3d sess. Senate. Rept. 1021) Serial no. 3148

Printed also as House report 1954 of the 53d Cong., 3d sess.

[Description of the organization, by J. W. Powell. Criticisms as to duplication, with recommendations for improvement. The report was made by C. W. Haskins and E. W. Sells, experts]

53 —— *President's commission on economy and efficiency.* Summary outline of organization of the . . . Geological survey. (*In its* Message of the President . . . on economy and efficiency . . . Washington, 1912. v. 1, p. 43-8)

54 —— *Geological survey.* Abstract of petitions, memorials, requests, and estimates relating to the operations of the hydrographic branch of the United States Geological survey. February 8, 1901 . . . Washington, Govt. print. off., 1901. 71 p. (56th Cong., 2d sess. Senate. Doc. no. 152) Serial no. 4039

55 —— —— Abstract of reports, letters, clippings, in-

dorsements, resolutions, petitions, and requests showing the demand for the topographical work of the United States Geological survey. January 21, 1902. . . . Washington, Govt. print. off., 1902. 67 p. (57th Cong., 1st sess. Senate. Doc. 136) Serial no. 4231

56 U. S. *Geological survey.* Annual report, 1st—1880- Washington, Govt. print. off., 1880-

[The first [preliminary] report described plan of organization and publications. Beginning with the second report, these annual reports contain the administrative reports of the Director, together with technical papers, these comprising several volumes. "Beginning with the Twenty-third, each annual report consists of one volume, such papers as had been included in the Director's report being thenceforth published as professional papers, bulletins, and water-supply papers."]

57 —— —— Building for the Geological survey. Letter from the Director submitting a report on the question of a building for the Survey. [Washington, Govt. print. off., 1906] 6 p. (59th Cong., 1st sess. House. Doc. no. 621)

[Shows that quarters occupied by the Survey are inadequate and urges a new building]

58 —— —— The classification of the public lands, by G. O. Smith [and others] Washington, Govt. print. off., 1913. 197 p. (Bulletin no. 537)

59 —— —— Coöperation between the United States and various states in topographic, hydrographic and geologic work. Washington, Govt. print. off., 1910. 32 p.

60 —— —— Division of mines and mining, United States Geological survey . . . Letter from the Director . . . transmitting a report on joint resolution (S. R. 205) to provide for a division of mines and mining in the . . . Survey. [Washington, Govt. print. off., 1898] 12 p. (55th Cong., 3d sess. Senate. Doc. no. 40) Serial no. 3778

[Division of mines and mining, or a bureau of mines, recommended in report]

61 U. S. *Geological Survey.* Fuels and structural materials. Letters from the Secretary of the interior, transmitting a copy of a letter from the Director of the Geological survey embodying a summary of the results obtained in the investigations under the survey of fuels and structural materials at the testing plants at St. Louis, etc. . . . [Washington, Govt. print. off., 1906] 14 p. (59th Cong., 1st sess. Senate. Doc. 214) Serial no. 4913

62 —— —— Nature and uses of the topographic maps of the United States Geological survey. [Washington, Govt. print. off., 1908] 4 p.

63 —— —— The State geological surveys of the United States, comp. under the direction of C. W. Hayes. Washington, Govt. print. off., 1911. 177 p. (Bulletin no. 465)

64 —— —— Statement of purposes and results of investigations and tests of fuels and structural material, showing bearing on conservation of mineral resources. May 4, 1908. Washington, Govt. print. off., 1908. 14 p. (60th Cong., 1st sess. Senate. Doc. 460) Serial no. 5265

65 —— —— Statement of the Director of the Geological survey regarding the desirability of a more extensive recognition of the mining industry by governmental inquiry. [Washington, Govt. print. off., 1908] 7 p.

[Favors establishment of a mining technology branch in the Geological survey, or an independent bureau of mining technology, to supplement the geologic work of the Survey in behalf of the mining industry]

66 —— —— Testing of fuels and structural materials for the United States. [Washington, Govt. print. off., 1907] 8 p.

67 —— —— The United States Geological survey: its origin, development, organization, and operations. Washington, Govt. print. off., 1904. 205 p. (Bulletin no. 227)

[Includes account of the United States survey prior to establishment of Geological survey]

68 U. S. *Geological Survey* Use of moneys received from sale of maps; etc., Geological survey. [Washington, print. off., 1904] 3 p. (58th Cong., 2d sess. House. Doc. no. 130) Serial no. 4671

—— [Handbooks of the Geological Survey]

69 Handbook for field geologists in Geological survey; by C. W. Hayes. Washington, Govt. print. off., 1908. 159 p.

70 Hydrographic manual of the United States Geological survey. Washington, Govt. print. off., 1904. 76 p. (Water-supply paper, no. 94)

71 Instructions relating to the work of the United States Geological survey, to take effect May 1, 1903. Washington, Govt. print. off., 1903. 131 p.

[A companion manual which relates to details of a less important and more temporary character than the subjects treated in "Regulations." Contains general instructions on organization, correspondence, library, etc., of the Survey with instructions for individual branches as to reports, notes, etc. Salary tables show rates of payment]

72 Manual of topographic methods, by Henry Gannett. Washington, Govt. print. off., 1906. 8 p. 8 pl. (Bulletin no. 307)

73 Methods of stream measurement. Washington, Govt. print. off., 1901. 51 p. (Water-supply paper, no. 56) Reprinted in 1902.

74 Regulations and instructions of the United States Geological survey. Washington, Govt. print. off., 1911. 164 p.

75 Rules for the preparation of manuscript and illustrations designed for publication by the United States Geological survey. By Thomas Hampson. January, 1888. [Washington, Govt. print. off., 1888] 24 p.

76 Suggestions to authors of papers submitted for publication by the United States Geological survey, with directions to typewriter operators, by George McLane

Wood. . . . 3d ed. Washington, Govt. print. off., 1916. 120 p.

77 Service bulletin, August, 1916; organization and investigations in progress. Washington, Govt. print. off., 1916. 45 p.

78 Topographic instructions of the United States Geological survey; ed. of 1915. Washington, Govt. print. off., 1915. 215, A-X, 217-228 p.

Unofficial Publications: Books and Pamphlets

79 Atkins, J. D. C. Speech of Hon. J. D. C. Atkins of Tennessee in the House of representatives, Feb. 10, 1879. Washington, 1879. 23 p.

[Relative to bill which would place the scientific surveys and surveys of the General land office under the control of the Coast survey; the abolition of the office of Surveyor-general and the creation of the office of Geological director]

80 Brown, Charles O. Present public land surveys. n. p. [1878?] 6 p.

81 Clarke, F. W. The chemical work of the United States Geological survey. (*In* International congress of applied chemistry, London, 1909. Seventh international congress of applied chemistry. London, 1910. Section 2, p. 146-61)

["Sir William Ramsay congratulated Prof. Clarke on the ingenious and instructive series of researches which have been pursued for so many years in his laboratory"]

82 Everhart, Elfrida. Geological survey. (*In her* Handbook of United States public documents. Minneapolis, 1910. p. 108-12)

[Brief history of the service; organization and scope of work; descriptive list of publications of the Survey]

83 Extracts from letters and notices of eminent scientific men and journals in Europe and America, commendatory of the United States Geological and geographical survey of the territories . . . under the direction of

F. V. Hayden. Washington, R. O. Polkinhorn, 1879. 42 p.

84 Fuller, M. L. Hydrologic work of U. S. Geological survey in eastern United States. (*In* International geographic congress. Report, 1904. Washington, 1905. p. 509-14)

[Work of the Survey which deals with underground waters may be grouped under six heads: Bibliographic, statistical, technical, legal, scientific and economic]

85 Geological survey. (*In* Cyclopedia of American government. New York, 1914. v. 2, p. 74-5)

86 Hayes, C. W. The relation of the federal government to the mining industry. (*In* American mining congress. Proceedings, 1905. Denver, 1906. p. 46-59)

[Includes a brief history of the U. S. Geological survey and some considerations of its relations to the mining industry]

87 Hewitt, Abram S. Our national inheritance, and how to enjoy it. Speech of Hon. Abram S. Hewitt of New York, delivered in the House of Representatives, Feb. 11, 1879. Washington, 1879. 16 p.

[An eloquent plea for the institution and continuance of an effective geological survey. Recommends consolidation of all the offices of surveyors-general into one, the abolition of the contract system and the appointment of a surveyor-general of the United States]

88 Hollister, G. P. Hydrographic work of U. S. Geological survey. (*In* International geographic congress. Report, 1904. Washington, 1905. p. 515-22)

[Summary of investigations showing determination of quantity, including the study of underground waters; determination of quality of water; publications]

89 International geological congress. Discussion sur le coloriage des cartes géologiques. (*In its* Compte rendu. Washington, 1891. Washington, 1893. p. 208-11)

[Mr. Powell opened the discussion by exhibiting charts displaying the color schemes recently adopted for the general geological maps about to be published by the U. S. Geological survey. This discussion is of importance because it helped to standardize the colors used in similar charts in other countries]

90 Marcou, Jules. The geological map of the United States and the United States Geological survey. Cambridge, Mass., The author, 1892. 56 p.

[Points out the incompetence and extravagance of the U. S. Geological survey and recommends a method of constructing and publishing the Geologic map of the United States]

91 National academy of sciences. Letter from O. C. Marsh, acting president, transmitting . . . report on the scientific surveys of the Territories . . . Dec. 2, 1878. [Washington, 1878] 5 p. ([U. S.] 45th Cong., 3d sess. Senate. Mis. doc. no. 9) Serial no. 1833

92 —— Report of National academy of sciences. Letter from O. C. Marsh, acting president, transmitting report of operations . . . during the past year. Washington, 1879. 25 p. ([U. S.] 46th Cong., 1st sess. House. Mis. doc. no. 7) Serial no. 1876

[Contains report of Special session of Nov. 6, 1878 on the surveys of the territories, and its Report on surveys of the territories. Appendix D]

93 —— Surveys of the territories . . . report. Dec. 3, 1878. [Washington, 1878] 27 p. ([U. S.] 45th Cong., 3d sess. House. Mis. doc. no. 5)
Serial no. 1861

[Includes correspondence between officials in the Department of the interior and the National academy of sciences]

94 Parker, E. W. Some experiments of the Geological survey coal-testing plant at St. Louis. (*In* American mining congress, Proceedings, 1905. Denver, 1906. p. 60-6)

95 Public lands survey. A consolidating and centralizing scheme proposed. Washington operators reaching out for more spoils. Major Powell to the front again in the role of chief consolidator. Argument against the

proposed change. Let our ornamental surveys be consolidated; but let our public land surveys alone. n.p., n.d. 17 p.

[Objections to "project to abolish the U. S. Surveyor-general's offices in all the states and territories and consolidate them and our public land surveys in, or under, one bureau . . . at the National capital."]

96 Singleton, Otho R. The Hayden survey. Speech of Hon. Otho R. Singleton, of Mississippi in the House of representatives, June 13, 1878. Washington, 1878. 7 p.

[Favors increased appropriation for the Hayden survey. Quotes extracts from letters from scientific men in this country and abroad in praise of Hayden's work]

97 Smith, George O. The possibilities and limitations of Geological survey work as applied to the mining industry. (*In* American mining congress. Proceedings, 1908. Denver, 1909. p. 138-46)

98 Surveying. By officers of the United States Geological survey. [New York] 1905. [419]-48 p. (International engineering congress, 1904. Paper. no. 29)

99 Van Tyne, C. H. and Leland, W. G. Office of the Geological survey. (*In their* Guide to the archives of the government of the United States. 2d ed. Washington, 1907. p. 227-9)

[Contains a statement of the correspondence of the service on file, a brief account of the Survey, its library, publications and sources of information]

100 Walcott, Charles D. Coöperation of the United States Geological survey and state mining bureaus. (*In* American mining congress. Proceedings, 1905. Denver, 1906. p. 82-98)

101 —— The United States Geological survey. (*In* Geological society of Washington, Washington, D. C. Presidential address. Washington, 1895. p. 3-24)

[Description of activities and progress of Survey up to Dec. 1, 1894, and extent of coöperation with states]

102 Walcott, Charles D. The work of the United States Geological survey in relation to the mineral resources of the Unted States. (*In* American institute of mining engineers, Transactions, 1901. New York, 1901. v.30:3-26)

[Recommends a Division of mines and mining in the Geological survey, to be changed later, if deemed desirable, to a Department of mines and mining. Shows why the Geological survey cannot undertake all the work that should be done for the mining industry]

103 Whitney, Josiah D. Geographical and geological surveys. From the North American review for July and October, 1875. Cambridge, Welch, Bigelow and company, 1875. 96 p.

104 Wigginton, P. D. Speech of Hon. P. D. Wigginton, of California, in the House of representatives, June 13, 1878. Washington, 1878. 14 p.

[On proposed amendment to reorganize that branch of the public service relating to the surveys of the public lands. Asserts that government may save $100,000 annually by consolidating sixteen offices into one]

Periodical Articles

105 The administration of the U. S. Geological survey. Science, Aug. 30, 1907, n. s. v.26:286-7.

106 Baker, Marcus. The U. S. Geological survey. Geographical journal, Sept. 1895, v.6 :252-60.

[A critical appreciation of the office in general and of J. W. Powell, director, in particular. Exposition of plan of work, its difficulties and accomplishments]

107 Branner, J. C. Policy of the United States Geological survey and its bearing upon science and education. Science, Dec. 7, 1906, n. s. v.24:722-8.

108 Brown, W. N. Propriety of government surveys for local use. Engineering record, July 17, 1915, v:72:789-90.

[Protest against permitting the Geological survey to establish a policy of undertaking detail surveys—particularly county surveys]

109 Burchard, E. F. Contributions of the United States Geological survey to architects. Journal of the American institute of architects, Feb. 1917, v.5:75-7.

110 Byrd, R. H. Measuring the flow of a stream; how water powers are accurately calculated. Scientific American, Mar. 22, 1913, v.108:268-9.

[Describes gauging cars used by the U. S. Geological survey and recommends that many additional gauging stations should be established]

111 Chapman, R. H. The United States Geological survey. Canadian mining institute. Journal, 1910, v.13:372-92.

[Descriptive and historical account]

112 Clarke, F. W. The chemical work of the U. S. Geological survey. Science, Aug. 6, 1909, n. s. v.30:161-71.

[Summary of results and description of different kinds of analyses]

113 Claudy, C. H. Model photographic laboratory. Scientific American, June 17, 1905, v.92:483-5.

[Photographic plant of the Geological survey described]

114 Dana, James D. Geological survey of the public domain. American journal of science and arts, Dec. 1879, 3d ser. v.18:492-5; Jan. 1880, v.19:78-81.

115 Early efforts of the government to aid mining were unsystematic. Mining congress journal, Nov. 1915, v.1 :620-2.

[Short account of early explorations and surveys]

116 English technical papers profuse in compliments [for the U. S. Bureau of mines and the Geological survey] Mining congress journal, Oct. 1915, v.1 :543-4.

117 General results achieved by the geological surveys in twenty-five years. Scientific American supplement, June 4, 1904, v.57:23770.

[Summary of Bulletin 227 of the United States Geological survey]

118 Geological survey and Bureau of mines assets in case of war. Mining congress journal, April, 1916, v.2 :161-4.

119 Geological survey makes effort to interest layman. Exhibits at Panama-Pacific exposition planned so as to appeal to general public—government spends $25,000. Mining congress journal, Mar. 1915, v.1 :121-2.

120 Geological survey maps now obtainable from postmasters. Mining congress journal, Oct. 1915, v.1 :555.

["Only postmasters in regions that have been recently mapped have been asked to handle the maps"]

121 Geological survey ready to aid in public work. Director Smith refutes charge that his bureau has infringed on private engineers. Mining congress journal, June, 1915, v.1 :264.

122 (Die) geologische landesdurchforschung der Vereinigten Staaten, während der letzten jahrzehnte. Globus, Dec. 10, 1908, bd. 94 :341-6.

[Reviews of the work of the U. S. Geological survey, with particular reference to its publications]

123 Hobbs, William H. One phase of Washington science. Science, Oct. 11, 1912, n. s. v.36 :477-9.

[Brief review of the presidential address before the Geological society of Washington, Dec. 13, 1911, re the relation of applied geology to that of the science of pure geology, as well as to the evolution of economic, political and social conditions]

124 Hoyt, J. C. Surface-water investigations; U. S. Geological survey. Engineering news, Dec. 23, 1915, v.74 :1213.

[Describes work of Surface-water division, water resources branch, U. S. Geological survey]

125 Johnson, Edwin F. Remarks upon the present mode of conducting land surveys in the United States. American journal of science and arts, Jan. 1831, v.19 :131-41.

[Outlines manner of organizing a system adapted to each state in the Union and urges need of such reform]

126 Marcou, Jules. The national surveys. Letter from an eminent French geologist—solutions of the questions before Congress—the letter of Mr. O. C. Marsh considered. Washington correspondence of the Cincinnati gazette, dated Jan. 29 [1879]

127 Mitchell, H. C. Competition of the U. S. government with private engineers. Engineering news, April 22, 1915, v.73:789-90.

[Letter written to Engineering news in protest against any engineering bureau of the United States government (including the Geological survey) entering into unfair competition with engineers in private practice. Reply of G. O. Smith, Director of the Survey, follows text of letter]

128 —— Competition of the U. S. Geological survey with private engineering practice. Engineering news, May 6, 1915, v.73:902-3.

[Second letter regarding unfair competition, following that published in Engineering news of April 22, 1915]

129 Morse, H. S. U. S. Geological survey and the topographical survey of Cincinnati. Engineering news, May 6, 1915, v.73:902.

[Letter from Director of Cincinnati Bureau of municipal research re "Competition of a government bureau with private engineering practice," and exonerating the Survey from such competition in the case of the topographical survey of Cincinnati]

130 New library [of the Geological survey] will contain everything on geological and mining subjects. Mining congress journal, Oct. 1916, v.2:437-8.

["With completion of new building of Department of interior space will be available for needed expansion in library of Geological survey. Miss Julia L. V. McCord, librarian, important factor in assembling collection."]

131 Newell, F. H. The United States Geological survey. Technology review, Oct. 1902, v.4:485-92.

[A brief appreciation of the value to the Geological survey of its comparative freedom from Congressional supervision and an explanation of the necessary qualifications for those desiring to enter the service as a career]

132 Norton, W. H. New list of topographic maps for use in high schools. School science and mathematics, Mar. 1913, v.13:236-40.

[To be used in studying land forms. List of 40 maps]

133 Packard, A. S. *jr.* [Note concerning the] new U. S. Geological survey in charge of Mr. Clarence King. American naturalist, May, 1879, v.13:343-5.

134 Sargent, R. H. Progress of Alaskan topographic surveys, by the U. S. Geological survey. American geographical society. Bulletin, July 1912, v. 44:481-92.

[Summarizes the work accomplished and describes methods of making surveys]

135 Smith, George O. Plain writing . . . address before American mining congress . . . Mining congress journal, Nov. 1915, v.1:579-81.

["Director Smith of Geological survey . . . tells of his ambition to have reports of his bureau in language of the people"]

136 —— The policy of the Geological survey. Science, Sept. 27, 1912, n. s., v.36:401-3.

[Answer to the statement of President Van Hise, in his paper "A national university" (printed in the issue of Science for Aug. 16, 1912) that the U. S. Geological survey "is almost exclusively a department of practical geology" and "is not contributing in any large way to the advancement of science"]

137 —— The possibilities and limitations of Geological survey work as applied to the mining industry. Mining and scientific press, Nov. 23, 1907, v.95:652-4.

138 —— Public need for applied science becomes more evident than ever. George Otis Smith in annual report on work of Geological survey points to increasing research work—Bureau enlarging policy of speeding up statistics. Mining congress journal, Jan. 1917, v. 3:19-20.

139 —— Relations of Geological survey to mining industry. Mining world, Nov. 23, 1907, v.27:924-6.

140 Smith, George O. The United States Geological survey and its relation to the United States Coast and geodetic survey. Science, May 12, 1916, n. s. v.43:659-65.

[Address delivered at Centennial exercises of the U. S. Coast and geodetic survey, April 5, 1916. An appreciation of what the Coast and geodetic survey has been to the Geological survey and an account of the coöperative work of the two surveys.]

141 —— The work of the U. S. Geological survey. Past and present work. . . . Plans and opportunities for the future. Engineering and mining journal, Nov. 30, 1907, v.84:1019-20.

142 —— The work of the United States Geological survey. Mining world, Jan. 21, 1911, v.34:136-7.

143 Smith, W. S. Administration of the U. S. Geological survey. Science, Aug. 30, 1907, n. s. v.26:286-7.

[Correspondence between the Director and former employee of the Survey who alleges arbitrary methods and commercial spirit in the Survey]

144 The U. S. Geological survey. Science, Feb. 22, 1907, n.s. v.25:309-11.

145 The United States Geological survey. Canadian Mining institute. Quarterly bulletin, April, 1910, no. 10, p. 109-29.
Published also in the Journal of the Canadian mining institute, v. 13, 1911, p. 372-92.

[Gives an outline of the methods of work employed by the U. S. Geological survey]

146 United States Geological survey at the Panama exposition. Science, Mar. 12, 1915, n.s. v.41:383-4.

[Description of its exhibit in the Palace of mines and metallurgy]

147 Uses popular language on topographical maps. Mining congress journal, June, 1916, v.2:274.

[Sheet just issued to be the forerunner of others in popular language]

148 Walcott, C. D. Principles which govern the Geological survey in its relations with other geological surveys and working geologists. Science, Nov. 30, 1906, n.s. v.24:692-3.

[Reply to Dr. Branner's charge that the Survey is an undemocratic organization which abuses its power to the disadvantage of state surveys and of individual geologists]

149 —— The United States Geological survey. Popular science monthly, Feb. 1895, v.46:479-98.

150 Western surveys. From *The Daily Tribune,* Denver, Colo. Jan. 23, 1879. 5 p.

151 What the government is doing to save our national resources. Scientific American supplement, July 25, 1908, v.66:62-3.

[Praises the Geological survey for its work in conserving the national resources. It has saved millions of dollars for the government]

Manuscripts

152 Gunn, Richards & Co., *New York.* Geological survey (*In its* Reports [on the] Department of the interior. Oct. 1, 1907. p. 126-9)

153 U. S. *Geological survey.* Memorial of Major J. W. Powell on the occasion of the presentation of his portrait to the survey. Mar. 24, 1905. Remarks by G. K. Gilbert, response of C. D. Walcott, remarks by W. H. Dall. Washington, 1905. 8 l.

[In the Library of the Geological survey]

154 U. S. *President's commission on economy and efficiency.* Detailed description of accounting records and procedure of the Geological survey. 107 l.

[Draft of report on Interior Department. Part I. Appendix VII]

155 —— —— History, organization, and activities of the Geological survey. 106 l.

156 —— —— Report to the President on the land survey services of the government. June, 1912. 106 l.

AUTHOR INDEX TO BIBLIOGRAPHY

[The arabic numbers refer to items, not to pages]

INDEX

Remote Depossession

Irene Hickman. D. O.

Cover design: Painting by Janet Heinicke
Chairman, Art Department, Simpson College
Indianola, Iowa.

REMOTE DEPOSSESSION

Published by
Hickman Systems
4 Woodland Lane
Kirksville, MO 63501 U.S.A.
(816)665-1836

ISBN 0-915689-08-1

TABLE OF CONTENTS

ACKNOWLEDGEMENTS

I am deeply grateful to the many people who by touching my life in myriad ways have led me to this work. I do wish to express a special thanks to those who have contributed case reports to be included here.

A special thanks to Suzanne Meister for her encouragement, friendship and her skilled editing of the manuscript.

I also wish to thank in advance all those who are yet to become involved in Remote Depossession.

FORWARD

If I had not known Irene Hickman for over thirty years, I doubt that I could have been persuaded to write this Forward. I am well aware of how preposterous her story must seem to the uninitiated reader who chanced to pick up this volume.

To put any credence at all in what she has to say one has to be willing, at a minimum to assume:

(a) that there is a non-physical part of the self which does not die when the physical body dies;

(b) that because of various circumstances associated with the person's death, this soul-entity in some instances fails to progress onward with its own further evolution but instead comes to reside in another physically living body (in other words, that "spirit possession" appears in some sense to be a real phenomenon);

(c) that this condition, which is generally disadvantagous to the unsuspecting host person, can sometimes be treated successfully through persuading the parasitic entity to move on "toward the Light"; and

(d) that it is possible for this to be accomplished remotely, without the conscious awareness of the host individual (although presumably with the knowledge and assent of that individual's "higher self").

Modern science has taken a dim view of the first proposition, although on rather shaky grounds. Pervading practically all of science is the assumption that its basic information inputs are essentially limited to the data of physical sense perception. That being the case, it is not surprising that evidence suggesting a non-physical but nonetheless real aspect to human experience tends to experience ostracism. Most practitioners of "hard" science would claim this is simply a matter of Occam's razor, or the principle of parsimony: Never introduce a new concept if the phenomena can be explained without it. However, recent work with near-death experiences, regression therapy, children's memories of past lives, and other related areas indicates that a more tolerant attitude may be forming.

The second and third propositions are supported by the work on "spirit releasement therapy" as reported in recent books by Edith Fiore (*The Unquiet Dead,* Dolphin Doubleday;

1987) and William Baldwin (*Spirit Releasement Therapy:* Human Potentials Foundation Press, Falls Church, Virginia; 1993), and the brief treatment of spirit possession in the impressive two-volume treatise *Regression Therapy: A handbook For Professionals* (Winifred Lucas, ed. Deep Forest Press, Crest Park, California; 1993).

It is the fourth proposition which comprises the novel contribution of Dr. Hickman. Having trained in spirit releasement therapy under William Baldwin, she became concerned about the difficulty of distinguishing between effects attributable to release of a possessing entity and effects due to the suggestions of the therapist. Thus she developed a method of remote treatment, such that the one being treated need not know of or consciously participate in the process. This book is an accounting of that novel treatment method.

Obviously, this work need to be replicated by many other therapists before the healing community will give much credence to it. At this point Dr. Hickman makes no claim other than an honest reporting. But because there will be a temptation to dismiss her report without a hearing, we need to comment on why it is not totally outrageous.

The idea that someone might be "possessed" by some discarnate being is one that many of us may have felt should have been left behind with the Middle Ages. True, there have been reports of possession-like phenomena in every culture and every era. True, "psychical researchers" around the turn of the century investigated many such cases and took the "spirit attachment" hypothesis seriously. But surely with our present degree of scientific sophistication we can dismiss the idea of spirit possession as a prescientific concept for explaining phenomena we would now interpret in more scientifically acceptable terms! Perhaps.

Harvard philosopher/scientist, William James, was much concerned about this kind of puzzle. He felt science had been unduly restrictive in adopting an epistemology of positivist empiricism (that is, whatever we know scientifically in known through empirical research in the physically measurable world). He proposed casting the scientific net more widely, with an epistemology of "radical empiricism."

In his *Essays in Radical Empiricism,* James defines his term thus: "To be radical, an empiricism must neither admit into its constructions any element that is not directly experienced, nor exclude from them any element that *is* directly experienced." James was urging that science should be based *only* on experience, and that nothing within the totality of human experience should be excluded from being a potential topic of scientific investigation. How widely he meant this to be applied is indicated by the fact that he was involved for 25 years with exploration of psychic phenomena, spiritism and religious experience. He was not convinced of the conclusion of enthusiasts in these areas, but he insisted that the experiences are appropriate data for a complete science.

James' "radical empiricism" admits data from the senses, and thus includes within its purview the experience of the physical world. But it also encompasses the broad spectrum of inner realities found within the subjective life of the person. It also implies that no phenomena are to be excluded from investigation solely on the basis that they "violate known scientific laws."

Looking at human experience through the eyes of James' "radical empiricism," we must be impressed by the diverse realms of reported experience which imply something like a nonmaterial self that Western science appears unable to accommodate. For example, out-of-body experience (experiencing one's center of consciousness and perception to be somewhere other than behind the eyes---for instance, as above, looking down on the physical body) and near-death experience are apparently far more common than we usually take them to be, and often lead to a conviction that the "I" is something different from the physical body. Research on childhood memories of past lives has turned up numerous cases where the vericicality of the memories was substantiated. While alternative hypotheses are possible, the data strongly suggests something like the Eastern concept of reincarnation.

Additional support for taking the spirit concept seriously comes from research on "mediumistic" communication purporting to come from persons no longer living. For several decades either side of the turn of the century a great number of competent scientists, particularly in England (but including William James and others in the U.S.) were interested in these communications which supposedly originated from the "spirits" of the deceased. Again, while alternative hypotheses are certainly possible, some of the cases were quite evidential in the sense that information showed up which was known to the deceased but not otherwise available to the "medium." In addition there are anthropological reports from other cultures, indicating that in their perceived realities "spirit" manifestation and possession are not all that unusual.

Thus it appears that spirit attachment, or "possession," may be a far more common condition that is commonly recognized. If so, if there are indeed intrusions into the human system by unwelcome non-physical parasites, then it is essential that people be made aware of it and that more practitioners, both in the mental health professions and in the clergy, be trained to use the appropriate techniques to relieve the condition. If, furthermore, the techniques of remote depossession prove to be effective, and others can be trained to use them, much human suffering can be relieved.

That's a lot of "if's". The jury is still out, undoubtedly. Meanwhile, at the very least, Dr. Hickman's book is a fascinating read and raises some most interesting questions.

Willis W. Harmon

January, 1994

INTRODUCTION

There is a growing awareness among health professionals that it has become imperative to look beyond the traditional views of main-line psychology and medicine.

With a rapidly expanding body of evidence supporting the concept of personal survival of bodily death, we must now look beyond the narrow views of behaviorist psychology and the chemical approach to health and disease.

This approach is not new. Alexis Carrel, M.D., discussed the need for better knowledge of man in 1935 in his book, *Man, the Unknown.*

> "......*Men cannot follow modern civilization along its present course, because it is degenerating. They have been fascinated by the beauty of the sciences of inert matter. They have not understood that their body and consciousness are subjected to natural laws, more obscure than, but as inexorable as, the laws of the sidereal world. Neither have they understood that they cannot transgress these laws without being punished.*
>
> *They must, therefore, learn the necessary relations of the cosmic universe, of their fellow men, of their inner selves and also of their tissues and their minds.*
>
> *Humanity's attention must turn from the machine and the world of inanimate matter to the body and the soul of man....."*

Dr. Carrel goes on to expose the tragic consequences resulting from our persistent ignorance about our basic nature. This book is not intended to provide the kind of proof customarily demanded by orthodox science. I have evidence to satisfy my requirement of proof.

The strict limitations of science preclude experimental study of human subjects. Since no experiment using human subjects can be repeated exactly, perhaps it will become necessary for science to postulate a new paradigm. Lacking this wider view, science excludes research that could reveal the true nature of the human mind, the nature of consciousness and answer the question of survival after physical death.

I am convinced that as humans we consist of an eternal indestructible soul-spirit living in a physical body. The present body is only one of many we have inhabited through the centuries and which we will subsequently inhabit until we learn our lessons sufficiently well to qualify for a permanent home in a non-physical Paradise.

This conviction arises out of my own memory---going all the way back to creation---and the reports of past-life memories of patients, trainees and colleagues These memories include past lives in other bodies, past deaths and vivid memories of the experience between one death and the next entry into a physical body through birth. I have treated this matter at length in my book, *Mind Probe---Hypnosis.*

This evidence has largely been gathered using hypnosis as the tool for exploration. Other evidence includes near-death, out-of-the-body, dream and meditation as well as mystical and psychedelic experiences.

Perhaps the largest single body of evidence is the work of Edgar Cayce, the American mystic who produced over ten million words during "readings" between 1900 and 1945 while deeply hypnotized. This material has been preserved by The Association for Research and Enlightenment in Virginia Beach, Virginia, and is now available on a CD Rom disk.

The material in this present book comprises a new approach to human problems that has proved to be effective when other methods of treatment have failed. It presumes that we each consist of an immortal soul-spirit dwelling in a physical body. When the body dies, the indwelling entity leaves and has choices to make. It may go to the resting, learning place called "The Light." It may choose to stay near the dead body's burial place. It may stay around a familiar place of object or may attach to a living person.

Those who fail to go to "The Light" miss the opportunity to review past experiences, to learn from them and to prepare for a new beginning again by rebirth into an infant body.

The natural laws spoken of by Dr. Alexis Carrel are particularly applicable in reference to our multiple lives, deaths and rebirths. The laws of Karma determine the effects of choices made. "Whatsoever a man soweth, that shall he also reap." Galations, 6:7. The result may be either pleasant or unpleasant. Karma need not be thought of as punishment only. Rather it is best thought of as the result of our own past choices or at times the result of a need to be persuaded to learn when we are resisting the learning.

I am convinced that all past experiences are permanently recorded in our memory and are retrievable using a variety of methods. All that is required to retrieve these early memories is the desire to recall them, simple techniques for quieting the usual mental "noise," looking within and asking appropriate questions. Hypnosis is an aid for such retrieval.

There is absolutely no need for anyone to be kept in ignorance of their immortality---in both directions, past and future---of the nature of death and all available options following the death of the physical body.

The universe has provided a place for our soul-spirit to go for a between-life sojourn. Unfortunately, few people have become aware while in the body that such a place exists. Few people have sought these answers and therefore have only a limited understanding of what the dying experience will be like. At the death of the body, many are surprised and confused by the sudden realization that they are not "dead" although their body has ceased to function. Some even fail to realize that their body has died. This confusion appears to be the chief reason for the so-called dead to remain Earthbound or to attach to a living person.

Those that do attach to a living person often have a profound effect on their host, and may cause physical, emotional, behavioral or mental changes. These changes can include serious illness, addiction to drugs or alcohol, personality changes, phobias, inability to succeed at a chosen career, suicidal thoughts or a myriad of other problems.

The idea that spirits of the so-called dead might be around us, attached to us or even interfering with our performance or behavior is felt by many to be repugnant, frightening or even terrifying. I was recently visited by a couple who were troubled and frightened because the wife's sister who had died some months earlier had appeared repeatedly in their apartment. My initial response to their story was, "Let's suppose that instead of your sister's dying, you had died. Then you wanted to go visit her. How would you feel if she acted the way you are acting? She is no more to be feared than she was when she was wearing her body. She is not there to frighten or confuse you. She apparently does not know the proper place to go. Dr. Edith Fiore's book, *The Unquiet Dead,* has the exact instructions to be given to those spirits who hang around familiar places or people instead of going to "The Light."

The couple recorded the Fiore instructions on tape and played the tape during the night. The "dead" sister did not appear again.

If this book does nothing more than help people to know what to expect at death, then it will have served a useful and very badly needed purpose. Each of us must face death. Knowing what it will be like and what choices are most appropriate when the body dies can save so much unnecessary pain and suffering.

Various theologies have told us what the after-life will be like. There seems to be little observable evidence to support common theology. Such ideas as that at death we enter into eternal rest, or wait patiently in our grave for the resurection, or go immediately to the feet of Jesus or perhaps to everlasting punishment in Hell are not substantiated by experience.

I have treated hundreds of my patients using non-directive hypnotherapy. When they are guided into past time to locate a cause of their illness or problem, they may regress to an event in their childhood in the present life. However, most of them go much further back to re-experience incidents in former lives often including vivid details of earlier deaths. Intense emotions are common. The descriptions they give of these earlier experiences have all been consistent with each other and with the philosophical comments contained in the Edgar Cayce "readings." My reasoning is that if these reports were not illustrative of the way the universe is actually organized, surely someone of the hundreds of cases would contradict, deny or refute. The patients that I have worked with tell me that at the end of a former life they can view their dead body from above; can choose to stay around loved ones or a beloved place, can move on to a beautiful place or may decide to attach to the body of a living person. I am both impatient and saddened that so few people are willing to test theological doctrine by actual experience.

A general assumption exists that a belief in reincarnation and karma is un-Christian. Until 543 A.D., when the Emperor Justinian attacked the reincarnation concept, it was a part of early Christian understanding. Origen, an early Christian writer, stated in his work, *De Principiis:*

> *"The soul has neither beginning nor end....Every soul comes to this world strengthened by the victories or weakened by the defects of its previous life. Its place in this world as a vessel appointed to honor or dishonor is determined by its previous merits or demerits. Its work in this world determines its place in the world which is to follow this."*

Chapter One

POSSESSION OR ATTACHMENT BY ENTITIES

My experience has led me to accept the concept that we can be and often are invaded or possessed by beings or entities in addition to our own soul-spirit. This book is intended to be a handbook for those who wish to learn to locate, identify, dialogue with and release those entities who might be in or around a living person for a variety of reasons. The ideal would be to find all such trespassers and send them to their proper place. This is a huge undertaking. As more trained workers enter this field, the sooner there will be hope of total success.

Edith Fiore, PhD, author of *The Unquiet Dead,* and William Baldwin, PhD, author of *Spirit Releasement Therapy, a Techniques Manual,* have been the leading educators and trainers in depossession and spirit releasement work. They have developed very effective techniques for recognizing, confronting and persuading attached or possessing spirits to leave their hosts and go to a better place.

In doing this work, Fiore, Baldwin and others have also found that some attachments are not those who formerly had a human body of their own, but came from a place of darkness and have assignments of causing pain, harm, confusion or disharmony. This work is related to exorcism in that the attached or possessing entities are removed. It differs in that we send them to a safe beneficial place rather than just casting them out to where they could return to the original host or to someone else.

The "dark" entities might be loosly classified as 'demons," but I have found very few who exhibit the traditional "demon" qualities. They seem more to be rather unbright, bratty underlings of some more powerful "dark" authority. One possibility is that they are merely the result of negative thinking. This book will have much to say about how to deal with both the Earthbound, those who once had their own bodies, and the "dark" ones.

Both Fiore and Baldwin emphasize working directly with the person who has the attachment. This method appears to be very effective. My own question about such a method is, "How great a part does suggestion play in producing the results?" It seems to me that the changes reported might possibly be due to hypnotic suggestion alone.

Because of this possibility, and because I prefer all possible assurance that I am doing what I think I am doing, I have chosen remote depossession as preferable. By this remote method, the one being treated need not know of or consciously participate in the process. We also do remotes with the person being scanned present with us. We *always* ask permission of the high self of the target person and proceed only after permission has been granted. If the target high self responds with either a "No" or a "Not now" we do not continue.

Transcripts of a few actual cases, to be included later, will illustrate the step-by-step procedure. This is easy to learn and rèlatively easy to do. However, I do not recommend this technique for the timid therapist. Boldness, assertiveness and fearlessness on the part of the hypnotherapist are assets for this work.

Chaplter Two

REQUIREMENTS FOR A REMOTE DEPOSSESSION

1. Trained Hypnotherapist.

To prepare for this work the therapist needs to be trained in both directive and non-directive techniques. Another requirement is a thorough understanding of the philosophy behind regression or past-life therapy. There also needs to be an awareness of the potential of the hypnotized person's "Inner Wisdom." Self confidence and self assertiveness are most helpful. The timid had best not become involved in depossession work.

2. Willing Hypnotized Subject.

Willingness and interest in this work are essential. We also require approval of the hypnotized person's "Inner Wisdom."

3. The Induction.

There are many suitable inductions to help the subject reach a satisfactory depth of hypnosis. Each therapist develops techniques with which they are comfortable and which for them produce satisfactory results.

I usually select a progressive relaxation beginning either at the top of the head or in the feet. I ask the subject to imagine a wave of relaxation slowly moving over their body allowing the substance of the wave to just soak in and pervade them totally.

I suggest also that the subject imagine that each inhaled breath is pure relaxation and allow this to produce waves throughout the body. I suggest that they also let each exhaled breath carry any tension or discomfort up and out. I emphasize the importance of allowing the tongue to lie loose and limp on the floor of their mouth. Before awakening the subject after the first long induction, I say, "In the future, if there is an occasion for us to work together, I will take your left hand in my left hand and count from one to twenty-one. If you are willing, by the time I reach twenty-one, you will be deeply relaxed and ready for the work we propose to do. You will remain in control. Your "Inner Wisdom" will guide your choices as to whether to accept or reject whatever I say to you----whether to answer my questions or to remain silent."

The second and all subsequent times I work with a subject, I use this post-hypnotic suggestion of the left hand to left hand and the twenty-one count. I then instruct the subject to go to the correct level of relaxation for the work we propose to do and indicate that they have reached this level by nodding their head.

I have found several advantages to the use of this post-hypnotic induction. The subject seems to find a degree of comfort and security from my holding their hand. I am able to guage the degree of relaxation as their left hand and arm grow limp and heavy.

The instructions about remaining in control serve the purpose of easing any anxiety the subject might have about losing control. Although there is plenty of evidance that control is never lost, many people still react as if this danger does exist.

4. Selecting the target to be scanned.

We may begin with a list of 2, 3, or 4 names and addresses of people who have requested remote depossession by mail, or whose names have been chosen by those present. At times the person for whom we are doing the remote is present in the room.

The "Inner Wisdom" of the hypnotized subject is consulted with a question such as, "Is it appropriate for you to do a remote scanning of _______, _______, or _______ at this time?" We continue only after being given approval. We may ask about several names at one time, or ask approval of each one individually moving to the second and subsequent names after the first remote depossession is completed.

5. Locating the target to be scanned.

The hypnotized person (scanner) is asked to send a part of their consciousness to an address and locate the target person. (If we do not know the exact location of the one to be scanned at the time we are working, but know their location at an earlier time, the hypnotized person can find them at the earlier time.) We have been successful going back in time as far as a month. It is likely that an even longer time could be utilized. The scanner recognizes when the target has been located and nods their head or tells that the target has been found.

6. Permission of the high self of target is requested.

7. After permission is granted: We instruct the hypnotized subject to scan thoroughly both the body of the target and also the periphery and report anything they find that appears to indicate a presence. Common comments may include: "There's a black area in the head and the right shoulder" or "There's something in the chest or throat or leg." Sometimes only one area is noted; more commonly, two or more.

8. Identifying the type of entity.

Speaking to the entity, "You who are causing the darkness in (target), come forward, speak through (scanner) without harming her/him in any way or give her/him your thoughts and let them speak for you and identify yourself. Do you have a name?"

Earthbounds, those who have lived in a physical, human body, usually say "Yes" or give a name. "Dark" entities usually remain silent, or may say something like, "Go away, I don't like you" or "No." One surprising response was, "No, I don't have a name and I want a name. I' ve worked hard for a name." The two types of entities, Earthbound or "dark," are treated differently.

9. The Earthbound.

Some of these don't know that they are dead, they are confused or feel lost. Questions are chosen to help them comprehend their true condition. They often need to be guided through their death experience in order to realize that though their body has died, they continue to exist or "live" in their non-physical state.

We ask if they had been told about "The Light"----the proper place to go after the body dies. Answers are varied. A few admit knowing about "The Light" but may have judged themselves unworthy because of their wickedness, their drunkenness or, in one case, a black man felt that they didn't let "his kind" in. Many say they do not know of "The Light."

When these entities are told that they are having a damaging or detrimental effect on their host and are asked if that is their intention, the usual response is to deny wanting to hurt the host, and they often express regret and apologize if they may have caused harm.

Additional information is given about "The Light" and the desirability of going there. This is a process of "selling" until they express an interest in actually going. At this point, a call is given to anyone in "The Light" who cares about the attached entity to come for them, greet them, show themselves to them, communicate with them and help them with their transition. The Earthbound is then told to look up and tell what they see. In most instances, they will see and recognize a relative, a friend or an angelic figure.

A few times, the entity has reported that nothing was seen when they looked up. This indicates that the Earthbound had attached to them either one or more other Earthbound spirits or may even be carrying a "dark" entity. When those that are attached in tandem are released by appropriate techniques, the first located entity can then move to "The Light." Some entities are reluctant to leave the host because their intention is to help, or their hesitancy may be due to fear of the unknown.

Those who come from "The Light" to greet the Earthbound ones are able to communicate with them and explain what their existence is like in "The Light" and help convince the sceptics.

Very rarely, those who come are impostors. If we suspect an impostor, we ask the Earthbound to look into their eyes and touch their hands. (The hands will be cold and the eyes will appear red or dark.) Any impostor who comes is ordered to leave. The impostors leave at once. None have hesitated to leave of the few I have encountered.

When an Earthbound entity expresses a willingness to leave their host and go to "The Light," we ask that before they leave they look around inside the host's body and around the periphery to find any other attached beings of similar nature that they can persuade to go to "The Light" with them. This may or may not be successful. Reluctant Earthbounds may have to be handled individually until all have gone to "The Light." Willing additional entities leave with the primary one. We say, "Go in peace with our love and blessing." We deal with reluctant entities later by repeating the procedure.

10. Filling the vacated space or spaces.

We call upon the healing angels to "come and fill all spaces with healing Light that were previously occupied by those who have departed."

11. The "dark" entities.

When there is no answer to the question, "Do you have a name?" or they give some aggressive response such as, "Go away," "I don't like you," or "None of your business," we know we have encountered an entity who never had their own human physical body. I may respond with "I'm not going away." "I'm not here to be liked" or 'I'm making it my business." We call at once for the Forces of St. Michael, the Archangel, to come, cast a net of Light around the entity and pull it snug. If the entity refuses to answer questions, we ask that the net be pulled tighter and tighter until prompt and truthful answers are forthcoming.

Typical questions comprising the removal procedure include: (a) "Who sent you to (host)?" They may respond with, "You know," "I'm not supposed to tell," "The big boss," "The evil one," etc. (b) "What is your purpose in (host)?" The answers are consistent in indicating that they are under instructions or orders to harm, disturb, confuse, cause failure or illness or even to kill. They fear severe punishment from their commanders if they tell the name or fail to carry out orders.

We proceed to undermine their commander by asking whether they had been promised a reward or promotion for doing a good job at their assignment. Many report having been promised. None report having been given the rewards.

When these "dark" entities are asked what they have been told about Light, the response is that they must avoid Light as it would harm, melt or destroy them. We then ask the Forces of St. Michael to send a shaft of Light right through the "dark" entity repeating as many times as seems to be needed. Typical questions comprising the removal procedure include: "Did that hurt? Did it melt you? Did it destroy you?" They agree that there was no harm. We tell them that they have been lied to and deceived. We then ask them if they like working for a liar and deceiver. This usually produces a response of displeasure.

We continue to question and comment: "How old was (host) when you joined?" "What were your instructions about remaining hidden or secret?" "Now that you have been found, you have failed to follow instructions. What happens to beings like you who fail to follow instructions?" Responses are very much the same from entity to entity. They tell of a special place of punishment, a pit, a horrible place which they know by having spent time there or by having heard about it. We remind them that because they have failed they are in danger of being sent to the place of punishment, but that they have another option---to go with the beings around them.

"Dark" entities are suspicious and may require continuing efforts to convince. One effective method is to ask them to communicate with St. Michael's Forces, some of whom were once "dark" destroyers who have been transformed. "What do they tell you?" "Have they found a better way?" The rescuers respond that some were indeed once "dark" and have found a much better existence.

At the point the "dark" entity indicates a willingness to go with their rescuers, we say, "Look deep within yourself to the very center of your being and tell me what you find." (They all report finding a spark, a flame or dot of Light.) "Keep watching that Light and tell me what happens." (It always grows larger and brighter.) "What is happening to your darkness?" (It all disappears.) We continue, "Now that you glow, you attract any other "dark" beings in or around (host's) body. Call them to you. Ask each one to look within to see what is there and watch it grow. Are all that are in or around (host's) body willing to go with you?" Usually, all are willing. When we do encounter one or more reluctant ones, we ask all the willing ones to "Go in peace with our love and blessings." "As you leave (host's) body, call out through the universe to all beings of your former "dark" nature of equal rank and stature, of lesser rank and stature, of greater rank and stature. Call out. Are you calling?" They always answer "Yes." "Are they coming?" Again a "Yes." "How many?" There may be only a few who come or there may be masses too great to count. "Ask each one who comes to look inside and find their true nature at their center." They are all expected to do this.

Again the healing angels are called to fill the recently vacated spaces with healing Light. This is followed by a repeat of the previous steps with any that failed to go with those who left willingly.

14. Rechecking.

To determine if any have been hiding or were left behind, the body is scanned again. Any entities found are treated as above until the body is found to be clear of entities.

15. Filling Target with Light.

The scanner is guided to find the Light of the Target's spirit at the center of their being and visualize it growing brighter and larger until the entire body is filled with Light. If any area fails to fill, we must go back and deal with any problem remaining. When the body is filled with Light, directions are given to let the Light expand in all directions at least arm's length to form a comforting, protecting, strengthening capsule or bubble around target.

16. Others to be scanned in this session.

In the event that approval has been given by the "Inner Wisdom" of the scanner for more than one remote effort, we proceed to the others. I personally feel that three depossessions at one sitting is the maximum---or 1-1/2 hours, whichever comes first. We usually do only one depossession per session. In about twenty-five percent of the time we do two.

17. Permission to feel different.

We request that the high self of the target let them know that the change they feel is all right and also help them adjust to their new feelings and urges.

18. Scanner is cleared.

Scanner is directed to find the Light of their own spirit and let it expand filling their body. If any area fails to fill it indicates the presence of an entity in the scanner. We deal with any entity found in much the same way as we do with the remote.

Chapter Three

CHARACTERISTIC EFFECTS OF ATTACHMENTS

1. Sudden changes in behavior ---- such as in eating, drinking or sleeping, depression, increase in anger, rebelliousness, fear or panic or suicidal tendencies.

2. Physical pain ---- typically not accompanied by usual laboratory or x-ray findings. This kind of pain also fails to respond to traditional medical treatments.

3. Serious illness of unknown cause ---- Persistent and non-responsive infections have been shown to be caused by attached Earthbound entities who suffered from or died of such an infection. When the entity left, so did the infection. Unusual inflammatory or metabolic disorders and blood dyscrasias have disappeared after depossession.

4. Loss of energy ---- sudden diminution of physical energy is highly suggestive of possession. A common result of depossession is a marked increase in energy, increased vitality and stamina.

5. Sudden onset of alcohol or drug abuse ---- In such cases the possessing entity is usually the one who did, in life, indulge in drugs or alcohol. When influenced to depart, the entity usually takes the urge to use with them.

6. Memory and concentration problems ---- When someone accustomed to easy learning and high grades in school abruptly changes to low performance and poor grades, it is safe to assume that the possessing entity is the one with the learning problem.

7. Inner voices ---- If these voices express a reaction or point of view at variance to the one hearing the voices, there is most likely an entity responsible.

8. Multiple Personality Disorder ---- Many of the multiples are probably possessions. If that is the case, the treatment of choice is depossession, which when completed, removes the alternate personalities. Careful evaluation may be needed to determine whether or not what appears to be a Multiple Personality Disorder is really a possession.

9. Repetitious nightmares ---- particularly if the nightmares contain elements of violence or a struggle against threatening forces.

How Do Entities Gain Entrance?

In order to possess a living person, it is necessary for that person to be vulnerable. A number of different conditions provide the opening for entry. Entry is available during times of unconsciousness for any reason----a blow to the head, anaesthetic, drunkenness or drug overdose. Other ways we open to possession are: exhaustion, emotional crises such as intense grief, anger, fear or guilt. Severe illness also lowers our resistance to possession. The presence of one entity seems also to open the door to additional spirits. Deceased loved ones may stay around after their death and attach to the living with the intention of being helpful.

Channeling, mediumship, out of the body experiences or automatic writing are open invitations to becoming possessed. I also know of a few cases in which being hypnotized appeared to provide the opening to the entity, but I tend to think that the entities were already present and the hypnosis only helped discover them.

Childhood abuse of any kind acts almost like a magnet attracting entities wishing to attach. Some entities report having joined during a difficult birth or even during gestation. Occasionally we find entities who report having been with their host in a previous life.

St. Michael, the Archangel

Paintings of Archangel Michael show him with his foot on the head of a demon. He carries a sword in his right hand poised as if to slay the one under his feet. His powerful pose is symbolic of his superiority over lesser beings. Bill Baldwin's performing a depossession is an impressive sight. When he encounters a difficult "dark" (demonic) being, he regularly calls on the "Mighty Warrier Angels" of St. Michael. The effect of these forces of St. Michael appears to be that of capturing, helping to convince "dark' beings of their true plight and nature and escorting them home to a special place in "The Light" designed just for them.

St. Michael's warrier forces are very helpful when we encounter "dark" entities. I have grown to think of some of them as former "dark" ones who have become reformed. These have lost all of their dark, troublesome nature and have become rescuers.

It seems appropriate to think of St. Michael as the head-master of a training camp or reform school for demons. Some who come as helpers communicate with possessing "dark" ones admitting that they were once "dark" and have been transformed into glowing beings. Whether or not this is the correct and complete understanding of the way it is, it serves as a useful metaphor for this depossession work when we have found or believe we have found one or more "dark" ones.

Archangel Michael has been a part of religious tradition for centuries. The Princeton Index of Religious Arts lists over 900 artistic depictions earlier than 1400 AD. Rudolph Steiner, born in Austria, and the founder of the Anthroposophical Society, made many references to Michael in his lectures and writings. One emphasis on the mission of Michael is to point the way for humans toward the Christ Being. In Jewish, Christian and Islamic mythology, Michael is given the role of Leader of the Jews, Prince of the Church and Guardian against the Devil. The ancient Persian religion of Zoroaster emphasizes the presence of good and evil forces. They designate the evil forces as "Abrimon, the "energy spirit," "demon of demons" or "Spiritual Foe."

Again and again sacred and secular writings describe an ongoing battle between the influence for good as against wickedness and evil designs. Rudolph Steiner speaks of the power of the Dragon working within man trying to drag him down.....but that man in spirit is the luminous angel whose task has always been the total defeat of the Dragon. This glowing figure, St. Michael, illumines and warms and provides the strength to conquer the Dragon's might in our lower nature. This dragging down does appear to be the purpose of those "dark" entities we have encountered. We do obtain help from the Forces of St. Michael. This adds weight to the concept that Michael has as his purpose the overcoming of the "Evil One."

Approximately one fourth of the healings of Jesus were the casting out of unclean spirits and demons. Although scripture makes no mention of sending those who were cast out to "The Light," it seems safe to assume that Jesus sent them to the proper place. He said, "These things that I do shall ye do and more." I feel that releasing attached spirits is included in that command. Some theologians claim to believe that that command pertained only to the twelve apostles. I prefer to accept it as being given to all of us.

Chapter Four

MY INVOLVEMENT IN THIS WORK

My life has been filled with a great deal of excitement and adventure. By far the greatest excitement of all of my experiences has come in the nearly five years since my depossession training. During these years I have been a part of or witnessed transformations in people as possessing entities were removed and sent to "The Light." It now appears to me that the possible transformations are limitless. Neither time nor space provide limits. Only the scarcity of trained workers prevents or delays widespread changes in the minds, the health, the emotional tone and behavior of great numbers of people. Let us dream about the possibilities----changes that this therapy can produce. Just as a little yeast (when conditions are favorable) grows and multiplies until it leavens the whole loaf, we could permeate the world.

Some of us may want to limit our efforts to immediate family, friends and business associates. Others may wish to begin depossessing national and international personalities. I have no doubt that many prominent newsmakers are influenced in their attitudes and actions by the presence of attached entities. If attached entities are triggering choices toward enmity, bigotry, hatred and violence, and if we are successful in releasing these entities to "The Light," there is reason to hope and expect that much of the ugliness, terrorism, racism and warfare could be eradicated. This is a sacred occupation. We cannot know how effective this approach will be until remote depossession techniques have been used by many trained people.

One of the cases that generated this kind of enthusiastic optimism dates back to the Spring of 1989. This remote was done on four young men newly-adjudged incurably criminally insane. The Doctors had evaluated them as hopelessly unfit for release into society and had expressed the opinion that they would need to remain behind bars for the rest of their lives. All four men were under 30 years of age. One day, I as the hypnotist, and Mona, a very good hypnotic subject, decided to explore the possibility of helping these young men. Mona, under hypnosis, sent a bit of her consciousness to the prison hospital, over one hundred miles away, located the young men one at a time, asked and was granted permission from each high self to scan them for the possible presence of entities.

We found and released entities, both Earthbound and "dark", from each of them. Their behavior changed at once and within a year all four had been reclassified and three had been released as cured.

Another remote depossession with prompt and gratifying results concerned R, a 19 year old in the Chicago area, who had begun heavy use of alcohol at the age of 15. This remote was done in my home in Kirksville, MO, about 400 miles from the boy's home. During the "remote" we found an Earthbound entity who stated that he had been driving drunk, had hit a tree, had died and had entered R when he was 15. We convinced him that there was a better place for him in "The Light." He left willingly with a beloved friend who came for him. R. stopped his drinking as abruptly as he had begun it four years earlier. Eighteen months have now passed and he remains alcohol free.

Very recently a guest in my home did a remote depossession on her 84 year-old father who had been experiencing severe pain in his lower back, knees and feet. We located and removed entities from all three areas. About an hour after the procedure, she phoned him. He reported being completely pain free. We also did a releasement on her cousin and freed an attached grandmother who had died when the cousin was 27. This cousin had exhibited a marked personality change at twenty-seven becoming more aloof and distant with people---much like the personality of the dead grandmother. When phoned, the cousin expressed much more warmth and caring than was usual. Surprisingly, she said, "I love you," which she had never said before to this cousin.

In May of 1992, for a news spot on local TV, we did a remote depossession on a young girl of 11 years who had recently become unmanageable, Just as we finished the procedure, from three miles away, the child's attitude and behavior reverted back to her former normally assertive nature. Her rebellion, belligerence and contrariness vanished and has not returned.

During a lecture-demonstration in south Texas in March of 1993, a member of the audience, R..., said, "I wonder if this is what is wrong with me. After an extreme emotional crisis two years ago, I developed asthma. Medication keeps it under control, but it is there all of the time." J..., present in the room, was one with whom I had previously worked. I knew her to be a very good hypnotic subject. J...agreed to do a scanning of R....from across the room. We found a large, "dark" entity in her chest. We dialogued with this "dark" one in R...'s chest and convinced it to go with the rescuers. When this one agreed to leave with the Forces of St. Michael, we asked that it take any others of similar nature with him. We were told that there were other "dark" ones in R...'s feet and knees. Those in the feet were willing to go, but the ones in the knees refused. We directed the one in the chest to take the willing ones and "Go in peace with our love and blessing."

We then confronted the "dark" entities in R...'s knees. At this point, J...'s face contorted to become gargoyle-like and she began hissing. It required more than half an hour to convince these entities to leave. Then we asked the healing angels to come and fill all the vacated spaces with healing "Light." We then asked R... whether she had been having any trouble in her feet and knees. She replied that they had been hurting "real bad." I asked her how they felt now.

She stood up, moved her knee and ankle joints in various ways and said, "The pain is gone." Then, in surprise, "The asthma is gone too!"

This work is extremely satisfying. The satisfaction arises first from the simplicity of the techniques, the ease in using them and the rapid production of results. Although fearlessness is an asset in dealing with the entities----particularly the more prominent "dark" ones----there is no need to be fearful of producing undesireable results in the one being scanned remotely. I doubt that there exists another therapy with less likelihood of adverse side effects than is to be found in remote depossession therapy. The remote depossessions we have done for people present in the same room, produced obvious changes at once. For more distant scanned targets, we may recieve only a brief verbal report or none at all. I admit feeling a measure of dissatisfaction because I do not receive full reports on each case we handle remotely.

All sense of dissatisfaction can be dispelled by the receipt of Case Reports that are detailed with a full background history, how the session was conducted together with the noted results are very gratifying.

One report, which is included with only minor editorial changes, has illustrated to me that the whole effort is worthwhile.

BACKGROUND INFORMATION:

Barbara Sonin, a hypnotherapist from New Jersey, was my demonstration subject during a one-day seminar on Past-Life Therapy and Remote Depossession during the convention of the International Association of Counselors and Therapists in April, 1993, in Edison, New Jersey.

Three weeks later, Barbara and another participant in the I.A.C.T. workshop, Joyce Collela, came to my home in Kirksville, Mo. for two days of intensive training and hands-on practice. Both women demonstrated a high-level natural propensity for the work.

Upon returning to New Jersey, Barbara related her new interest in depossession work to a hypnotherapist colleague. This friend, S....T...., requested that Barbara hypnotize her to scan her troubled daughter, B... The following letter is from the mother, S.... T.... .

Dear Dr. Hickman, May 31, 1993

"My friend and colleague, Barbara Sonin, asked me on your behalf, to write you about my daughter, B.... and report on the recent depossession work we did on her. The change has been so amazing and profound, it has certainly expanded my beliefs about all this!

"Allow me to give you some background first of all. B.... is an eight and one-half year old adopted child. She was "given" to us by a teenager from New York who had heard about our spiritual community in Tennessee that took unwanted children. Her mother was seven and one-half months pregnant when she started on a Greyhound Bus journey from upstate New York. She got as far as Buffalo, collapsing on the bus from dehydration and a urinary tract infection. The bus driver took her to a nearby hospital. She was admitted and put on I.V.s and antibiotics for a few days. After she recovered, she set off on her journey again and arrived "at our doorstep" in remarkably good condition. As she was only sixteen when she got pregnant, had dropped out of school and had moved away from home at fourteen, chances are her nutrition and pre-natal care were pretty poor. We had excellent midwives and prenatal care on the community (The Farm). By the time her delivery date came, she had a beautiful, normal, easy vaginal birth. My husband and I were both there at the mother's request----it was a beautiful experience for all of us.

"When the baby was four months old, we moved to Los Angeles, California, so that my husband could set up a west coast market for his food business, which had begun as a cottage industry on the farm, (Farm Foods). That winter B.... got pneumonia and almost required hospitalization. Thank goodness, she didn't have to stay, but there was one long and worrisome day at the hospital during which I carried her sick and feverish little body in and out of various hallways and rooms to get her chest x-rayed. It was at that time she picked up one of the spirits that we subsequently released during our session.

"She was always a very thin, tall, waif-like child----very affectionate, sensitive and psychic. At three and one-half, she was mildly molested sexually (fondling and kissing) by three neighborhood boys who took her into the woods behind our house. I think she picked up another entity at that time. When she was four, she spoke to me of angels and told me about two past lives she remembered. She was always very gentle and caring towards animals. We've had many pets. She tells of seeing those we have lost in spirit form.

"By kindergarten, she started to exhibit some mild learning problems and a slight immaturity compared to her peers. On the first day of first grade, (age six) she fell off the monkey bars of the swing set in our back yard and shattered her left humerus. We were right there and whisked her to the hospital where she was admitted for a total of four days. After an overnight elevation of the arm, they worked on her for three and one-half hours in surgery, piecing and wiring together the broken bones. She was a model patient and recovered beautifully. Even though the growth plates were disturbed, she has full use of that arm today. It is growing normally and she is a strong competive swimmer.

"That year, she began to have a lot of problems in school. I finally asked them to test her for a learning disability. She was tested at the beginning of second grade. She was diagnosed as having ADD (Attention Deficit Disorder), a learning disability and classified as perceptually impaired. She has been in special education at her school part time, and mainstreamed part time. She is taking Ritalin for the ADD, but the last two years have been very difficult in many ways.

"Her behavior has been confrontive, stubborn, overly-sensative and emotional. Finally, when she wouldn't allow anyone in our family to touch her or even brush against her, we decided in December to take her to a very helpful psychologist. He was also working with my husband and me on appropriate parenting for ADHD children and proper discipline for these notoriously difficult children at the time we did the depossession work on her. In fact we were in the midst of designing a sort of reward-punishment program for her.

"Barbara Sonin and I attended the Unlimited Humans conference together and heard you speak. Then she went to your Post-Conference Workshop on depossession. (I was too skeptical), When she returned from visiting you in Kirksville and told me of her experiences and what she had learned, I couldn't help thinking about my daughter and couldn't dismiss the possibility any longer that she might have some attached entities.

"One afternoon, Barbara led me through your spirit releasement technique. This was the closest experience I have had to "channeling" as these entities spoke through me. There was a blind boy that had attached himself to B....'s eyes and wanted assurance that he would be able to see before he willingly left with loved ones waiting for him in "The Light."

"There was Theresa who cried the saddest tears through my eyes----so afraid, so young----murdered at two or so. She had joined my daughter in the hospital when she was getting that chest X-ray during pneumonia. Theresa happily, easily left to join her mother in "The Light."

"There were several entities in B....'s stomach. One, the leader, was so angry and stubborn. I was surprised to find myself gritting my teeth and growling in obstinate anger as he spoke through me. This was a man who had been physically abused and literally kicked around all of his life. He said his name was Dave, but no one ever called him that. He had been called nasty, unkind names. When asked if he had heard about "The Light," he said, "Yeah, but I don't believe that shit." He was also reluctant and unwilling to go to "The Light," not believing that there was anyone who ever loved him there. Barbara perceptively recognized that he had an entity or two within him. We exorcised them accordingly---(A surly, nasty force that required help from the Forces of St. Michael and much encouragement.) Finally Dave was able to see an aunt he recognized in "The Light," the only person who had ever been nice to him in his life. Finally he left. During our dialogue, Dave said that upon dying in the hospital, he was frantic and panicked on finding himself out of his body. He floated around the hospital until he came across B.... as she was having her arm surgery. He commented that she was the sweetest thing he had ever seen and boom---that was it. He was inside her soft little tummy.

"The interesting thing here is that after B....'s surgery, she started gaining weight at an amazing rate. She had had quite a flabby overhanging stomach for the last two years that has concerned us greatly. I have been worried about the possibility of a developing eating disorder and asked the doctor about it. She was compulsive at times about food and our attempts to curtail the excessive appetite had all failed. Since we released these several entities from her stomach---the area of greatest congestion---she has lost eight pounds in less than a month and has grown one-half inch taller. This occurred literally overnight. She came home from school the day we did the depossession as happy and joyful as she had been in early childhood and has remained that way ever since.

"I perceived a little flurry of negativity about four or five days after our initial session and told Barbara I thought there was something else there. We did one more session on her and found a "dark" one way down on the floor of her pelvis, hiding. It had been attached to Dave, and had hidden when Dave left for "The Light" hoping we wouldn't find him. St. Michael and his angels easily convinced this one to go and my daughter has been a sweet, delightful pleasure to be with every day since then. Situations that had been extremely difficult in the past and had led to lengthy, stubborn arguments and subsequent punishments are a breeze now. She is flexible, agreeable, relaxed and so much happier.

"I must add that there was one spirit residing in her solar plexus that we encountered in our initial session named Anias. She was very, very lethargic, had been sick for so long and I feel that she was drugged. Her energy felt extremely slow and laid-back in my body and mind as she spoke through me. Her doctor reached out to her from "The Light." She had been such a hypochondriac in life that he was the only one she trusted to tell her about "The Light." She took hold of his hands to go into "The Light" as she left my daughter.

"I was a little concerned about the call I had to make to the psychologist the week after we did all this to cancel our appointment with him. I knew I wanted to tell him the truth and wanted to be careful how I selected my words so that he wouldn't think I was completely crazy. I told him I had done some remote healing work on her, that it had had such a profound effect that I didn't think we would be needing his services any more. I called it "Spirit Releasement." When he asked me to describe it, I did my best, using terms like "Light" and "darkness" etc. He was wonderful, understanding, delighted that this had worked, and asked if this wasn't kind of like Shamanic soul-retrieval. What a relief! I loved him for it.

"Many thanks to you for ultimately teaching this to Barbara so that she could teach me. I dedicate myself to this work now. I use it in my hypnotherapy sessions---I can't believe I dare to bring it up---and it is being well received. I have also been doing this on my own in my daily meditations. I can usually "see" the person I'm working on quite clearly and have released an awful lot of unwanted entities into "The Light" in the last three or four weeks. I have also read *The Unquiet Dead,* by Dr. Edith Fiore. In speaking to a couple of friends, I learned that they have been aware of this type of healing work a lot longer than I have. I feel like I have learned so very much this month.

"But the greatest proof of all this is my happy, sweet, joyful daughter. Finally, she can get on with her life of developing as a normal eight and one-half year old girl. She still has problems, still is not the most intelligent kid in her school class---but everything is so much more easily workable now. We weather through as things come up quite smoothly. I simply surround her with white Light often. She is shining and thriving, thanks to this glorious work."

"Thank you."

Respectfully yours,

S.......... T.........

Certified Hypnotherapist.

Chapter Five

WHERE DO THE 'DARK' ONES COME FROM?

NEGATIVE THOUGHT FORMS ---- FALLEN ANGELS ----MINIONS OF SATAN.

There are several theories. There may be some truth in each of them. The theory that negative thoughts and emotions give rise to beings having a continuing existence after their creation, would answer the question, "Why are there so many "dark" beings?" If each bit of pessimism, despondency, vindictiveness, bigotry, hatred, malicious gossip and desire for revenge creates one of these beings their numbers are great enough to keep us busy rescuing them for a very long time.

If they are all fallen angels, it is logical to expect that there could be a finite number. Perhaps completing the job of rescue and reform is forseeable.

Instead, perhaps these are creations of a Satan, A Prince of Liars, the Evil One, who was cast out of heaven with his angels. A common view is that this Evil One has as his purpose the taking over of the Earth and its population. It seems reasonable to me that this one also has creative power, albeit the power of darkness, to create beings who are as slaves or servants to do whatever they are told to do. These beings need not be intelligent, only obedient.

The responses we get when we discover and dialogue with these "dark" beings indicate a kind of hierarchy similar to that of a military organization. Orders from a commander must be followed. Orders may be passed on to those of lower rank and stature. There is an expectation of promotion or reward when an assignment is well completed. (These rewards are seldom given.) All orders or assignments are to be followed exactly without question. "Dark" ones are either led to believe that Light does not exist or is to be avoided because it is dangerous, will burn or destroy. They admit when questioned that they have been instructed to remain silent and hidden. They are threatened with severe punishment for failure to do their job or for being found. They fear both the Light, the threat of punishment and their commanders.

Whatever their origin, they appear to exist in great numbers. I am inclined at this time to accept all three theories as responsible for producing some "dark" beings. Another concept is that the presence of one "dark" entity attached to a person generates the negative thinking that creates additional "dark" ones. This is an area for continued exploration and research. However, such research cannot be made to fit the usual "science paradigm." Remote depossession as has been outlined and described here is very effective. As other workers become involved in this work, I would expect that even more successful techniques will be developed.

Is This Exorcism?

This question can be answered both "yes" and "no." It is exorcism in that the effect is to detect, dialogue with and release attached entities from their host. This method differs from the usual exorcism ritual in that the attached entity is not merely cast out of the host. This would be a little like removing nails from a flat tire and throwing them into the street where the original tire or a different one could pick them up again.

Remote depossession as described in this book detects entities, classifies them as to their basic nature, educates them and convinces them to go to a proper place from which they will not return to cause further trouble. This may be a battle with a Prince of Darkness we are fighting. I find it pleasant to think of the victories we are achieving without violence or destructiveness. Instead, we are capturing the "dark" entities and leading them to the kind of transformation that changes them to become helper-rescuers. How disconcerting it must be for the Lord of Darkness to observe his slaves, servants or warriors being lured away and trained to turn their energies toward his defeat. How long might it be before we capture all the "dark" ones?

We find that we have help when we call for it. Usually a few rescuers from St. Michael's realm bring a net of Light to inclose and contain the "dark" ones so that we can educate them. In difficult situations, I have seen Dr. Bill Baldwin call on many saints, Mary, Jesus and even the spirits of whales and dolphins.

The vast number of "dark" ones suggests to me that they are being continually created in the dark realms and have been sent to cause all kinds of pain, trouble and difficulty in order to gain control of the world.

It seems to me that we have allowed these "dark" beings to gain a larger measure of influence or control by failing to recognize their presence and intent. Even many of the theological "leaders" have discarded the idea of "evil." They seem to have forgotten that Jesus cast out many "unclean spirits" and "demons." He advised us to do as he did. 2000 years later, we are just getting started. Although scripture makes no mention of the place these beings were sent, it seems safe to assume that Jesus could send them directly to "The Light."

Chapter Six

VARIETIES OF CASES

There is wide variation in the kinds of problems we have treated by the method of remote depossession. Each person's problems are unique and must be handled with care and sensitivity.

Attached entities are able to produce physical discomfort or malfunction, can disturb relationships, interfere with successful lives, retard development in children and can produce or enhance mental or emotional illnesses. The following cases have been chosen to show this wide variation in the kinds of problems in which remote depossession has been beneficial.

A Child Going Blind

This girl, aged eleven was losing her eyesight because her eyeballs were growing soft. Traditional medical treatment had had no beneficial effect. We found that a part of the problem was from a former life in which she was sacrificed at age eleven in a ritual fire ceremony. This memory was contributing to the problem. Also attached to her were a number of spirits of other children who had also been sacrificed in the same way. We freed the attached children to "The Light" and communicated with the high self of the girl with the softening eyes that she need not react to the fire in this life---that the eyes had been burned in another life. Almost immediately after this procedure her eyes began to improve.

The Dying Professor

In March of 1992, a friend who teaches at Northeast Missouri State University in Kirksville where I live, asked that we do a "remote" for a colleague, Ed. C., who was critically ill.

The story she told us was that Ed had developed breathing problems so severe that he had required mechanical assistance for several weeks. Her information included the doctor's inability to determine the cause of the breathing distress, or find any helpful therapy. The available information indicated that the doctors were unable to help Ed return to health and that death could be expected within days.

Scott Herner, of Scandia, Minnesota, was visiting me the day of this request. He had been the instrument for quite a number of "remotes" and readily consented to checking out Ed. Scott was hypnotized, his "Inner Wisdom" was consulted and the proposed scanning of Ed was approved. Scott sent a portion of his consciousness to the University Hospital in Columbia, Missouri, ninety miles south, and located Ed. Scott was able to describe Ed's appearance accurately.

We asked permission of Ed's high self to scan him for the possible presence of entities. Permission was granted.

Scott scanned Ed's body and found several entities. The dominant one was located in Ed's chest---a man who had died of emphysema and had joined Ed more than fifteen years earlier. This entity's presence had gradually weakened Ed's lungs resulting in the increased breathing difficulty. When Ed became so ill that he was admitted to the hospital, he was invaded by three additional Earthbounds who had recently died in that hospital. One was a girl whose death had resulted from a drug overdose. The other two were victims of other illnesses. The entry of these three entities caused a marked increase in the severity of Ed's symptoms.

All four of the entities present were rather easily persuaded to leave Ed and go to "The Light." We then asked the healing angels to fill the vacated spaces in Ed with healing Light.

Within less than an hour from the time we did this depossession, Ed began to improve. Within two months he was back at work. His health continues to be good.

Ed has given me permission to use his name in telling of this story. Although I have full permission, I have chosen to use the correct first name and an initial to spare him some of the derision from probable sceptics who might read this book. When Scott and Ed met face to face for the first time in August of 1993, there was immediate recognition by Scott.

The Nurse With Painful Shoulders

One of my nephews from California visited me during the summer of 1991. He had never been hypnotized before, but when I told him what I had been doing with remote depossesion, he asked me to hypnotize him to see if we could find the reason for persistent pain in his wife's shoulders that had defied all attempts both to diagnose and treat it. He located his wife, a school nurse, two thousand miles aeway, found and removed entities from her shoulders. The pain left at about the hour we did the procedure and did not recur.

The Boy and His Crayons

A boy of five in Ohio, who had been near death from spinal meningitis at age eighteen months, was scanned remotely from Missouri. After his severe illness he had seemed to develop normally in every way except one. He was in nursery school where the use of pencils and crayons was common. This child, however, would only run the pencil or crayon in and out through his fingers and had never made a mark on paper. His grandmother was visiting me. She, under hypnosis, sent a piece of her consciousness seven hundred miles, scanned her grandson, located and freed a feeble-minded child who had died in the same hospital when this boy was ill and had attached to him. The next week, the grandson was voted artist of the week at the nursery school. This could have been a coincidence or it could have been the result of our work. Unfortunately there is no certain way to determine this. I choose to think that our work was responsible for the change.

The Naughty Girl

Another grandmother requested a remote depossession of her six-year-old granddaughter who had misbehaved so badly that she was no longer permitted to ride the school bus and was at the point of suspension from school for refusal to follow rules. When she was sent to the principal's office, she would kick and scratch the principal. This was one of the more difficult cases, but after two sessions with remote techniques with two different scanners, she showed a definite change toward being happy and self-controlled.

The Mental Hospital Patient

A friend requested a remote depossession for her mother who was in a mental hospital in New York state and had deteriorated to the point that she seemed not to remember her daughters and refused to talk to them. After the procedure, during which we found and removed several entities, both Earthbound and demonic, she improved remarkably and was being considered for discharge

The Difficult Supervisor of Nursing

Depossession has also been helpful in improving troubled relationships. A nurse I know in Manila told me of having a supervisor who had at first been very kind and helpful, then suddenly had become quite the opposite. We did a remote scanning of the supervisor and found two Earthbound attached entities (Relatives who had died at about the time of her change of manner.) We released these two to "The Light" and the supervisor returned again to her former kindness of manner.

It appears that there are no limits to what can be done by this method. It is exciting to think about all the possibilities. Perhaps we can reverse patterns in people who are troubled, criminal, ill or those falling short of their potential by removing entities that were interfering.

Chapter Seven

QUESTIONS AND ANSWERS

IS SPIRIT ATTACHMENT REAL?

The concept of entity attachment or possession is repugnant to many people. Such thought must of necessity include the idea of death---an idea so unpleasant that it is practically taboo. In recent years, research studying people who have had a near-death experience has produced a great body of evidence completely compatible with itself as to the the nature of the dying experience. Ever growing numbers of reports of people, including children, who have been pronounced "dead" and then returned to life supports the concept that though the body appears to be "dead," there is a non-physical (entity) part that can and does function outside of and apart from the body. This non-physical self can hear, see, smell, think, wish and manifest increased awareness while outside their body. These functions do not require a brain and invalidates much of the "right-brain---left brain" concept that is so popular.

Popularity of an idea does not give it value and validity. Refusal to accept a valid concept does not destroy its value.

Either we survive bodily death or we do not. Our survival is not dependent upon anyone's belief. Either spirit possession does take place or it doesn't. Truth is not dependent upon acceptance. If spirit attachment is real, then it is important that we learn about every aspect, learn how to detect and properly remove any intruding or trespassing entities.

Failure to understand this possibility greatly limits our helpfulness in dealing with maladies of various degrees of severity which are caused or complicated by the presence of trespassing entities. Proper understanding of possession and the use of appropriate remedial techniques---including remote depossession---has the potential of producing world-wide transformation.

One effect that attached entities have on us is that they appear to make our decisions for us. It is my understanding that whenever we allow someone else, living, dead or demonic, to make our choices, we are abdicating and thereby weakening our choice-making equipment.

A good way to decrease our vulnerability to entity attachment is to consistently make all of our own decisions, being willing to accept the consequences of any wrong choice we may make. Entities are unlikely to attach to someone who will not be controlled.

HOW LONG DOES IT TAKE TO LEARN REMOTE DEPOSSESSION?

I can only compare this question to a similar one---How long does it take to learn to drive a car, or cook a meal or play the piano?

Individual talent and application levels are so varied that there can be no rigid answer to such a question. Among the ones I know who are doing this work, there are a few with only minimal training but much natural talent and previous training and experience with hypnosis.

I had been involved in hypnotherapy since 1950 when I received training in spirit releasement in May of 1989 during three 3-day workshops in my home in a period of eleven days. These workshops were taught by Dr. Bill Baldwin. I began using the techniques cautiously and found them rather easy to follow and the success level high both with the remote process and when dealing directly with an afflicted person.

As with any skill, both practice and additional training sharpens our ability and builds our confidence. There will be some who will begin working at remote depossession just by reading this book. The techniques are simple enough that for a few, no further training will be required. However, I recommend that anyone involved get any additional training available.

Ideally the depossession therapist would be thoroughly trained and experienced in directive hypnotherapy, non-directive hypnotherapy and past-life therapy as well as spirit releasement or depossession.

HOW BADLY IS THIS WORK NEEDED?

We do not know of a certainty how many people's lives are complicated---even destroyed---by the effects of attached entities. Dr. Edith Fiore, author of *The Unquiet Dead,* has stated that she estimates that about seventy percent of those who seek her psychological counseling have attached entities. Other estimates go higher.

I would expect to find entities in 99% of those addicted to drugs or alcohol and nearly as high a percentage for the tobacco addicts. It is probable that "dark" entities are contributing to many if not all criminal or violent acts. It is likely that most, possibly all, victims of childhood or other abuse may have attachments. The perpetrators of the abuse are also probably hosts to "dark" entities.

This adds up to a huge number of people who could and would benefit from remote depossession---since many people, who need to be freed from the effects of attached entities, do not seek help. There are still so few people trained in this work that even those who do seek help find it difficult to find someone who is doing spirit releasement either directly for the troubled one or by remote techniques.

Whenever I lecture or give training workshops on this topic, the requests for help far outstrips the time we have to treat them. There are also frequent phone or mail requests for remote depossession. The urgency I feel to train as many workers as possible is compelling. I cannot even imagine that there could be a surplus of workers doing remote depossession work.

WHO ARE THE FORCES OF ST. MICHAEL?

This question, like several others gives rise to speculative answers. One possibility is that some are of angelic origin whose duty it is to assist St. Michael. It appears that many of these Forces are former "dark" entities who have been reformed (or transformed) from troublemakers to helpers.

When I am confronting a "dark" entity, I tell them to communicate with the rescuers, asking of their past nature and present existence. The answers are always consistent. Some of the rescuers will report that they were once "dark" beings working by orders of an evil boss or commander, that they were rescued and have been trained to come to rescue other "dark" entities. Some even admit to being very bad in the past until they learned a better way. They may add that they no longer feel fear of punishment.

Asking "dark" attached entities to question those who have inclosed them in a net of Light has been very helpful. Many "dark" beings have been skeptical or suspicious of my words, but seem more willing to believe or accept the statements of St. Michael's troops.

CAN RELEASED ENTITIES RETURN TO THEIR HOST?

The possibility is present. Anything CAN happen. However, it is most unlikely that they would choose to do so. The Earthbound entities who go to "The Light" find it far more pleasant than staying attached to a living person or hanging around a place or object. Once they go to this wonderful place, there is no desire to return except for brief visits from time to time.

Neither do "dark" entities choose to return. Once they are freed from their commanders and fear of the pit---the place of punishment---they would hardly select to return to that former status. Once training is complete they are able to join others of St. Michael's Forces to help rescue other "dark" ones.

Traditional exorcism emphasizes only the attached entity's departure from the host, but does not send these entities to a place where they will want to stay. Entities just "cast out" could most likely reenter their former host or the body of someone else.

IS REMOTE DEPOSSESSION THERAPY DANGEROUS?

This question too has various answers. Some workers in this field feel that a special ritual filling the room with the image or visualization of Light together with special prayers is the prerequisite to doing a spirit releasement.

I see this differently. It seems to me that such ritual is chosen by those who expect some sort of danger or trouble. This to me is negative thinking. We tend to experience whatever we expect.

I do not expect trouble. So far there hasn't been any trouble. I am certain that the power of Light is much stronger than any "dark" power. I am convinced that the "dark" entities are functioning out of fear and that we humans have dominion over them. They *must* do as we say, but at times persuasion is required before they fully realize our dominion. This may require being assertive---even aggressive at first.

As each of the "dark" ones begin to make the choice to go with the rescuers, I treat them with gentleness and compassion. Whereas their opening comments in our dialogue may include rudeness, vulgarity or profanity, they always lose this behavior during the session and often thank me for helping them.

Earthbound entities usually go willingly as soon as they realize that they have the choice. Only one has indicated a desire to harm their host---a slave in a former life was seeking revenge.

As for any danger to the person being scanned, I see no possibility. There could conceivably be times when the attached entity failed to leave---though this has not happened in the cases I know about. There are absolutely no undesirable side effects of remote depossession.

I usually have Rosary prayer beads present, not to avoid trouble, but to keep connected with our primary rescuer, Lilith/Eve/Mary, who gave us the Rosary.

WHAT IS LIFE ALL ABOUT?

The statements in this section are basically from material produced by Edgar Cayce while he was in deep hypnosis. I have modified it somewhat based on my own experience.

We were all created at the same time on the sixth day of creation to be companions and co-creators with Diety. We were given a bit of divine creative power and the free will allowing us to choose how that creative power would be used.

Very early we began to misuse this creative power and to forget our purpose. All of these events are a part of our permanent memory record at a deep level. We remember existing in Paradise without bodies, remember our first error, our first pride and rebellion. Our "Inner Wisdom" also remembers and longs to return. It is our "Inner Wisdom" that will lead us back again if we listen to it.

Ten billion souls became so disturbing, prideful and rebellious that it became necessary for us to be banished to the equivalent of "Devil's Island"---the Earth. It is my understanding that the banishment was intended to be permanent. We experimented with Earth things, including animals, plants and stones. We also invented hybrid creatures such as those who were half human with the other half bird, fish or animal. We were making no progress toward returning home again.

The first of the twin souls created---Amilius/Lilith---volunteered to come and rescue us. They have continued in this effort to the present time and will continue until all of us have returned home. They have helped to establish every religion which worships one God. Familiar names which they have worn include Adam and Eve, Enoch, Melchisedec, Joseph, Joshua, Jeshua and Jesus and Mary. The Amilius/Lilith pair are still exerting their efforts to bring us all back home again.

One of the reasons that discarnate spirits should go to "the Light" instead of attaching to living folks is that in "The Light" there is training to move us along the path to right choices. We might think of the between-life time as the lecture course with the Earth life in a human body as the laboratory where we can demonstrate what we have learned or what we have failed to learn. We repeat the failed lessons until we can earn a passing grade. There seems to be no time limit for this learning. We can stall as long as we wish. All of us will earn our graduation in time.

There seems to be two required courses for graduation plus a great many electives. The first required course is "acceptance" of whatever is without resentment, bitterness, anger or rebellion. The second is to say to whatever Diety we understand, "Here am I. What do You want me to do? THY WILL BE DONE in and through me." When we have accomplished this plus our chosen electives, we are ready to return home to continue the purpose for which we were created---companions and co-creators with Diety.

This work with spirit releasement and remote depossession is a sacred work to move all of us nearer to our permanent home in Paradise.

A universe that operates on these principles depicts a Diety that I find more worthy of the title "God" than the one I was taught about in Sunday School. I was taught that we are all sinners because Adam and Eve ate the forbidden apple and that the only way out of this sinfulness is to "believe" in and accept a personal savior. This seemed unfair to me even when I was very young. I was taught that I was to love God, but since he seemed to be so unfair to me and my family---poverty, illness and death of a mother when I was ten and one-half years old---it was not easy to so love.

The concepts of reincarnation and karma have helped me to understand the reasons for things. Learning that I had had two recent lives of wealth---wealth that I did not appreciate or share---has helped me to understand the need to experience poverty in this life. We are here on schoolhouse Earth to learn whatever lessons we have failed to learn in earlier lives.

CAN REINCARNATION BE PROVED?

This question can only be answered on an individual basis. In spite of the ever-growing mass of supporting evidence through the work of Brian Weiss, M.D., Robert Jarmon, M.D., Raymond Moody, M.D. and others, there remain many people who choose to ignore all evidence that appears to contradict their personal belief system.

The only convincing experience that provides what may be called "proof" is the remembering or reliving past-life experiences under hypnosis. There is usually not only vivid recall of events, but the re-experiencing of all emotions relating to the past-life happening. If these reports were only fantasy, the emotion would be absent.

Once someone has relived an event from a former life and has felt and released all the past emotions, there remains for that person no further doubt that the experience was real.

Not only can we experience past lives, but we can also experience past deaths. This is a most effective way to help people understand that death is not the end of existence. Although the body dies, the recent inhabitant of that body, the entity, is very much "alive" and can describe being up above the dead body "watching it float down the river" or "watching the tiger eat me" etc. They report that they may stay around their dead body or some familiar place or object or may travel any distance with complete freedom. After one or more of such experiences, all fear of death is replaced by a certain "knowing" that the real self is non-physical and eternal.

At the end of each life, for those who go to "The Light," there is a review and evaluation of lessons learned and unlearned. The realization of incompleteness is accompanied by the awareness that they must return again to Earth to finish the task or reach a higher level of perfection.

When we attain sufficient awareness of the meaning of our existence, we then qualify for "graduation" from Earth and are no longer required to return in an Earth body, but may choose to return as a teacher or leader. If I were naming any who might have returned by choice after their "graduation," I would include the names of Mahatma Ghandi, Albert Schweitzer and Mother Teresa. There are likely many others who taught and served without becoming so well known.

WHAT KINDS OF RESISTANCES DO WE ENCOUNTER?

Earthbound entities give a variety of reasons for failing to go to "The Light." One who had died in a fire said that he thought "The Light" was more fire; another said it was too bright and that he turned away. Other reasons given include "I'm not good enough to go there."; "I have to stay here to help..."; "They don't let black folks in there."; "There's my father and he wasn't good to me before."; "I've heard about "The Light" but I don't believe it."; "I couldn't be dead. I'm right here."

We deal with each resistance as it appears. I often feel that selling used cars would be similar. We continue to sell the idea until the entity is willing to risk testing what we tell them. Once they test and find our words to be valid, they move on in the right direction.

"Dark" entities appear to have very limited choices. They have been sent and have been given instructions by a being of darkness who has greater stature than themselves. They must obey their orders or be punished. They are motivated almost entirely by fear of being sent to the place of punishment---the Pit. A few have claimed to take pleasure in causing pain or trouble.

The majority are rather easily persuaded to go with the rescuers from St. Michael. A few are very resistant. Resistance may take the form of moaning, snarling, vulgarities, cursing, constant interrupting, expressions of disbelief and challenges to what we tell them. One even used laughter as a resistance ploy, scoffing at mere mortals who he claimed were incapable of understanding "dark" beings, their origin or purpose.

For the very obstinate "dark" beings, we ask the Forces of St. Michael to take the entity that they have surrounded with the net of Light to the place of punishment, suspend them over the edge and refresh their memory of what it is like. The typical response is, "I don't want to go back there again!" After this experience they become more willing to listen and pay attention to what we are telling them about other options open to them.

During my first few months of dealing with attached entities, there were a few occasions during which the entity with whom I was dialoguing seemed to suddenly vanish. Some "dark" entities seem able to enter and leave the body of their host at will. Now, as soon as I suspect the presence of a "dark" one, I call for help. I request the Forces of St. Michael to bring a net of Light and cast it about the "dark" one. I ask that the net be pulled snug and slowly but steadily tightened until I receive prompt and truthful answers from this one. A few "dark" entities have expressed displeasure or discomfort or have asked to be let out, but none have escaped when I use this technique.

Persistence is the main requirement in dealing with resistance.

WHAT IS HYPNOSIS?

Hypnosis is commonly thought of as a state in which the hypnotist has control of the mind of the hypnotized. There is a natural aversion to being controlled that prevents many people from experiencing the pleasure and benefit of being hypnotized.

Control of the mind is not a factor in the hypnosis experience. The hypnotized subject is able at all times to respond or not to any suggestions or request. This is because all hypnosis is really self-hypnosis. The hypnotist or hypnotherapist has the role of a teacher. The subject is free to follow or not. The relationship can be likened to that between a music teacher and pupil. The teacher can instruct, but cannot force the pupil to follow instructions.

However, because of the misconception concerning "control" of the mind persists, I always include in any session the following statement, "You are in control. You are free to accept or reject anything I say. You are free to answer my questions or to ignore them and remain silent. Your "Inner Wisdom" will guide your choices so that they will be right for you." I also empnasize that those I train should make the same statement.

There are many definitions of hypnosis---as many as there are those who define it. I prefer to think of the state of hypnosis as one in which we are in a closer communication with our "Inner Wisdom," a part of our mind that has all the answers we need. In the state of hypnosis we also move higher on the suggestibility scale. It seems to me that we are somewhere on the suggestibility scale all of the time. Techniques of hypnosis tend to move us to a higher point.

Hypnosis is characterized in most cases by a profound relaxation of the body and a quieting of the mind. In such a state we can probe the mind to determine its potential. There is so much still to learn and discover during this kind of exploration. I think of this probing as a great adventure limited only by our ability to think of questions to ask.

HYPNOSIS---The tool for locating, processing and relieving buried emotions so that their impact on our health, our behavior or our creativity is eliminated.

HYPNOSIS---A tool for relieving pain, easing stress, aiding memory, improving athletic skills, developing talents and reaching our highest potential.

HYPNOSIS---A natural, normal state that anyone can learn to experience, and which like other skills, improves with repetition.

HYPNOSIS---A gift of God to help us live more abundantly.

Everyone can be hypnotized, but some learn more quickly than others to go to deeper levels.

WHAT IS THE RELATIONSHIP BETWEEN THE ETERNAL SELF AND THE BODY?

The soul-spirit entity, the part that never dies, enters the body of the foetus usually about midway through gestation. According to Edgar Cayce, there are occasions when the entry may be as late as one month after birth. Within the soul-spirit are seven vortices or revolving wheels of energy called chakras. These seven energy centers attach to the seven endocrine gland systems.

The glandular centers are arranged as follows:

1. The Gonads---that part that produces the sperm or egg.

2. The cells of Leyden---that part of the gonads that produce the male or female hormones.

3. The Adrenals---that produce adrenaline and cortisone.

4. The Thymus---that influences growth and the immune system.

5. The Thyroid---that produces thyroxine or the "spark plug for energy."

6 The Pineal---the center for inspiration and intuition.

7. The Pituitary or Master Gland.

The entering soul-spirit brings with it all of the memories from all previous lives and between life experiences since creation. Memories are permanent and cannot be erased. The emotions attached to the memories can be called up, re-experienced and discarded. If not fully released these old emotions are able to cause various physical, mental and relationship problems.

Our old selfish sexuality is stored at the first center. Our fanaticism is stored at the second. Hates and fears are stored at the adrenal level. Non-sexual feelings of love or lack of it are stored at the thymus center. Life energy, or kundalini originates at the first center and rises up through all the others. When an obstruction from an old emotion is present, the flow is impeded. In the absence of blocking or obstruction, the life energy flows all the way through the seven centers.

The fifth center, the thyroid, is called the will center. Here we face the choices of "I want things my way!" versus "Thy will be done." This can be a point of blocking. If the energy can flow freely with enough force past the five lower centers, there is no further limitation or interference. The two higher centers, pineal and pituitary function to provide intuitive abilities, inspiration and balance to the entire being.

Another way of thinking of the connection between body and soul-spirit is that the soul-spirit controls the glands and the glands control the body.

Old emotions interfere with normal function and need to be located and expressed. The myriad kinds of healing that are only add-on influences are rarely permanent because they fail to seek, find and release the causal emotions. This is discussed more fully in my book, *Mind Probe----Hypnosis.*

When the body dies, the non-physical self, the entity, leaves the body carrying with it all of the memories and emotions. These memories and emotions are potent influences for the choice options of the entity. Vows such as "I will never leave you," or "I will get even if it is the last thing I do." also influence the choices of the newly dead. A thorough understanding of these basic principles gives a solid foundation for depossession work.

We who work with depossession are cautioned to avoid the temptation to think first that all problems are due to the influence of attached entities. It seems wise to first determine whether the problems are self-chosen without any external or internal buried influences. These influences can come from choices made either in the present life or in a past life.

After seeking causal factors in present-life choices or past-life traumas without finding all of the answers, it is safe to assume that one or more entities----either Earthbound or "dark" (demonic) may be responsible for existing problems. Many of the conditions that have responded well to remote depossession therapy were illnesses or problems that had not responded to traditional treatments.

There is much research still to be done to study the human mind. It is exciting to anticipate what may be the next discovery.

Chapter Eight

SAMPLE REMOTE DEPOSSESSION SESSION CHICAGO, ILL, MARCH 1993

The hypnotized subject was Rhona, a participant in a two-day seminar. I, Dr. Irene Hickman, conducted the session. We were asked to scan M...'s mother, J...., in Paris France, at a specific address. J.... was located by sending a bit of Rhona's consciousness to the address in Paris. Permission to scan J... was requested from J ..'s high self. Permission was granted somewhat reluctantly after we explained that our purpose was to help J.. . and intended no harm.

Dr. H. Begin scanning J.... Tell me what you find.

Rhona There is darkness in both arms and the solar plexus area.

Dr. H. Anything else?

Rhona Behind her in her aura.

Dr. H. What else?

Rhona Something jumping up and down on the top of her head.

Dr. H. Anything else?

Rhona That's all.

Dr. H. I call now to you in the arms of J..., M....'s mother to come forward, speak through Rhona without harming her in any way and identify yourself. Do you have a name?

(Asking for the name is intended to determine whether this is one who formerly had their own human body or is from the darkness.)

Rhona Pierre.

Dr. H. What happened to your body, Pierre? (The response was so slow in coming that it raised doubts as to this entities true nature.) Did you really have your own human body?

Pierre Yes.

Dr. H. Tell me what happened to it..

Pierre I drank some poison from a cup.

Dr. H. Then what happened to your body?

Pierre My body is there on the floor.

Dr. H. Then what did you do?

Pierre I just waited until someone came along.

Dr. H. When someone came along, what happened?

Pierre This person. She was kind.

Dr. H. How old was J....when you joined her?

Pierre Eight.

Dr. H. Do you know that you are having a detrimental effect on J... and that you are holding back your own development? How do you feel about that?

Pierre There's nothing else.

Dr. H. If there is a better place, are you interested?

Pierre I suppose so.

Dr. H. There is a place called "The Light"----a place for us to go when our body dies. It is more pleasant than staying attached to J.... In this place you grow and learn and eventually come back in a new body of your own. Would you like to have your own body again?

Pierre Yes.

Dr. H. I call now on anyone in "The Light" who cares about Pierre to come for him, greet him, communicate with him and help him with his transition. Pierre, look up and tell me what you see.

Pierre I see my mama.

Dr. H. Communicate with her. Ask her what it's like there. What does she tell you?

Pierre She says it's a good place.

Dr. H. Do you think that might be better than staying with J....?

Pierre Yes.

Dr. H. Then get ready to join your mama. But first I ask something of you. I would like you to look around inside J...'s body and around the periphery for any other beings that are lost and confused and are attached to J....'s body as you have been. You have a chance to gather them up and take them with you to this good place. Call them to you. Are they coming?

Pierre The one in the other arm is coming.

Dr. H. Are the others coming?

Pierre The one in the abdomen is coming.

Dr. H. And the one jumping up and down on the top of J...'s head. Is that one of your nature or of a different nature?

Pierre That one's not like me.

Dr. H. We will come back and check that one later. Take your mother's hand and the hands of those who are willing to go with you. Go in peace with our love and blessing. Thank you so much, Pierre, for helping. Now I ask the healing angels to fill with healing Light the areas just vacated by those who have departed. Now you, who are jumping up and down on J...."s head, come forward, speak through Rhona without harming her in any way and identify yourself. Do you have a name?

Entity No.

Dr. H. Have you ever had a name?

Entity No.

Dr. H. Have you ever had your own human physical body?

Entity No.

Dr. H. I call now on the Forces of St. Michael to come with a net of Light and wrap it around all of the "dark" beings in or around the body of J.... Pull the net snug and keep tightening it slowly but steadily until these beings respond to my questions promptly and truthfully. I call the spokesperson for the "dark" ones to come forward and answer my questions. Speak through Rhona without harming her in any way, or give her your thoughts and let her speak for you. What is your purpose there in J.... ?

Entity To make her miserable.

Dr. H. You've done a good job of that haven't you?

Entity Yes.

Dr. H. How old was J.... when you joined her?

Entity Thirty.

Dr. H. Were you given an order to make her miserable?

Entity No.

Dr. H. You thought that up all by yourself? I want a truthful answer. (no response) Do you take orders from any other being? Were you given any instructions before you came to J.... ?

Entity Hmmmm.

Dr. H. When you were given instructions. What were you told about the Light?

Entity I was told to avoid it.

Dr. H. What reason were you given for avoiding the Light?

Entity It would make me disappear.

Dr. H. That isn't true and I will show you. I ask now that the Forces of St. Michael send a shaft of Light right through this "dark" one---and another, and another and another. Did you disappear? Did it hurt you?

Entity No. No.

Dr. H. You have been lied to. You have been deceived. The one who gave you your orders was wrong. Are you willing to continue taking instructions and orders from one who lies and deceives?

Entity I would like to try that other thing.

Dr. H. You've heard about that other place? Have you been watching what we've been doing here today?

Entity Yes.

Dr. H. What do you think about what we have been doing?

Entity It's dangerous.

Dr. H. Dangerous in what way?

Entity I might get in trouble.

Dr. H. You were told to stay hidden too weren't you?

Entity Umhumm.

Dr. H. Now you've been found. What happens to beings like you who don't stay hidden?

Entity It's not good.

Dr. H. Have you been to that place of punishment before?

Entity Yes.

Dr. H. What kind of place is it?

Entity Pain, cold, lonely.

Dr. H. Now that you have been found, there is a chance that you will be sent there again. Would you like that?

> (Although some "dark" entities deny ever having been to the place of punishment---the Pit---they all admit knowing about it and that it is most unpleasant. Fear of punishment seems to be the primary motivation for following orders of their bosses.)

Entity No. It's bad.

Dr. H. Fortunately you do have another option. Those beings that have you inclosed in the net have come to rescue you and take you home. Some of them were once like you.

Look at them now. How do they look.

Entity They're bright.

Dr. H. How would you like to be more like them? (no response) You may communicate with them. Ask them if they ever had the job of causing pain and misery. Ask them what their existence is like now.

Entity They say it's good.

Dr. H. And they don't have to worry about being sent to that awful place. Do you think you might like to go with them?

Entity Yes.

Dr. H. In order to get ready to go, there is something you must do. Look deep within the center of your being and tell me what you find there.

Entity A little speck of light.

Dr. H. Watch that little speck. What happens as you watch?

Entity It's getting bigger and brighter. (tone of surprise.)

Dr. H. Keep watching. Now what's happening?

Entity It keeps growing.

Dr. H. That bit of light was there all the time, but you didn't know it. All it takes for it to grow is for you to recognize it. What's happening to your darkness?

Entity It's gone.

Dr. H. The others in and around J.... have been watching what has happened to you. Tell them to also look to the centers of their being. Tell me what they are finding at their centers.

Entity They are changing too.

Dr. H. You were all operating from fear. Now you have another way to function. Are all of the former "dark" ones ready to go now?

Entity Yes.

Dr. H. Fine. Go in peace with our love and blessing. As you leave J...'s body, call out through the universe for all beings of your former "dark" nature to come and follow. Call to all of equal rank and stature, of lesser rank and stature and of greater rank and stature. Even call to your commander. That one might come too. Call out. Are you calling?

Entity Yes.

Dr. H. Are they coming?

Entity Yes.

Dr. H. How many?

Entity 319.

Dr. H. Ask each one to look inside and find their true nature at their core. Thank you for helping. I now call on the healing angels to come and fill with healing Light all the spaces in and around J...'s body that were vacated by those beings who just left.

Rhona, once again scan J... Are there any other areas where there is need for work of this kind?

Rhona Around her ankles.

Dr. H. You in J....'s ankles, come forward, speak through Rhona without harming her in any way and identify yourself. Do you have a name?

Entity Lois.

Dr. H. Tell me about yourself, Lois. With a name you must have had your own physical body. What happened to that body?

Lois Hit by a car. I fell out of the door.

Dr. H. How old were you when this happened

Lois Ten.

Dr. H. Then what happened? The car hit your body, but what happened to you?

Lois There was a lady there.

Dr. H. Did you know this lady?

Lois I didn't know what to do. I went with her.

Dr. H. This is not your body. You're in someone else's body. There is a much better place for you to be. It's called "The Light"----would you like to know about that place?

Lois I think so.

Dr. H. I call now on anyone in "The Light" who cares about Lois to come for her, greet her, show themselves to her, communicate with her and help her with her transition. Lois, look up and tell me what you see.

Lois My sister.

Dr. H. I would like you to ask your sister what it's like where she is. What does she tell you?

Lois It's nice.

Dr. H. Do you think it might be better to go with your sister than to stay in this woman's feet?

Lois Yes.

Dr. H. Prepare to go, but before you do, please look around to see if there are any others who are attached as you have been because they did not know about "The Light."

Lois No others.

Dr. H. Fine. Then go in peace with our love and blessing. Goodbye, Lois. Again I ask the healing angels to fill the space vacated by Lois. Rhona, scan J... aagain. Are there more?

Rhona	No.
Dr. H.	Fine. Now look in the very center of J...., find the brightest point of her spirit, focus on it and watch it expand to fill her body. Tell me when her body is completely filled with Light or if you encounter any areas that fail to fill.
Rhona	She's filled
Dr. H.	Now let this Light expand beyond her body at least arm's length in all directions until she is completely inclosed in a bubble or cocoon of Light comforting, strengthening and protecting her. Tell me when this is complete.
Rhona	It's complete.
Dr. H.	Then bring that bit of your consciousness back from Paris, back to Chicago, to this room and to your body.
Rhona	There's something on my left shoulder.
Dr. H.	You on Rhona's left shoulder, come forward, speak through Rhona's voice without harming her in any way and identify yourself. Do you have a name?
Entity	Samuel.
Dr. H.	Samuel, what are you doing here on Rhona's shoulder?
S.	I was with J... I didn't go. (This is the only time this has happened in my experience.)
Dr. H.	You must have had a good reason for not wanting to go. What is the reason?
S.	I just didn't want to go.
Dr. H.	You may have missed a wonderful opportunity. Those who come from "The Light" to greet their loved ones all say it's wonderful there. Is it wonderful where you are?
S.	It's not easy here.
Dr. H.	Are you telling me that you can't stick to Rhona very easily?
S.	Right.
Dr. H	I call now on anyone in "The Light" who cares about Samuel to come for him, greet him, communicate with him and help him with his transition. Samuel, look up. Tell me what you see.
S.	My girlfriend, Louise.
Dr. H.	Do you want to stay here on Rhona's shoulder or do you want to go with Louise?
S.	I'll go.
Dr. H.	That wasn't a very difficult decision, was it?
S.	No.

Dr. H. Then go in peace with our love and blessing. Join Louise. I ask now for the healing angels to come and fill the space with Light that was just vacated by Samuel. Rhona, is there anything else we need to do for you?

Rhona No more.

Dr. H. Then focus on the center of your own being and find the brightest point of your spirit. Let this brightness fill your body and extend beyond forming a great cocoon of Light around you.

Rhona It's formed.

Dr. H. Prepare to waken. The forces within your body will equalize. I will soon waken you by counting backward from twenty-one to one using alternate figures. In the future, if there is an occasion for us to work together, I will take your left hand in my left hand and count from one to twenty-one. If you are willing, by the time I reach twenty-one you will be deeply relaxed and ready to go to work. If you are not willing, neither I nor anyone else will be able to induce this state in you.

You will always remain in control whether I or someone else is working with you. You are always free to accept or reject any and all suggestions that are given you. Your choices will be guided by your "Inner Wisdom" so that they are best for you. Ready, 21, 19, 17, take two deep breaths, 15, 13, 11, 9, more and more alert now, 7, 5, 3, 2, 1 (Clap of hands).

This was a new experience for Rhona. In response to questions, she said that scanning J....was like seeing her out in front and being able to see inside and through her. Rhona also expressed a feeling of increased inner energy and a deep sense of satisfaction.

Most people who are involved in this procedure report the same lift and harmony. Only a few have mentioned being tired or depleted afterward. I expect that we will find a way to enable all participants to feel pleasure and satisfaction rather than any other feeling when doing this work.

Chapter Nine

MORE CASE REPORTS

In June of 1993, I did a three-day seminar as a Buddhist retreat in California. During the time I was there, three staff members individually expressed their concern for a young man, Joe G., age 26, who had come to the center in great distress. He had told them that while studying for his Ph.D. he had volunteered to participate in some sort of experiment, the nature of which was not disclosed, and had felt himself to be "under attack" since.

He told them that he was hearing voices, feels he doesn't know who he is and feels almost dead. He was so frightened that he refused to leave his room. His food had been taken to him.

On Monday morning following the seminar, the three staff members had requested private sessions with me. C... was first. With her we did a remote scanning of another person she had sought to help and also helped her with a personal problem of her own. She told me that Joe G. had agreed to have a private session with me when she was finished. I suggested, "Let's fix him before he gets here." She agreed and under hypnosis again, scanned Joe G. (with his high self's permission, of course). We found and released several entities, both Earthbound and "dark."

A few moments later, Joe G. entered. All three of the concerned staff people were present as he came into the room. Joe appeared to be relaxed and calm. They all remarked about his changed appearance and manner. We just talked a few minutes during which time he said he felt good and seemed to be O.K. After he left the room, he walked about the retreat, talking to one of the children and appeared to be completely free of his problems

In August of 1993, I and an associate, Becky Marascalco, attended the National Guild of Hypnotists Convention in Nashua, New Hampshire. While there we did several private sessions. Becky is an excellent instrument for doing remote depossession. One person we worked with was particularly troubled. I shall call her Penny, (not her real name). Five weeks later we received a letter from her giving both the background and the after-effects of our work. Here is her letter:

"When I was a small child, six and under, I remember having a loving, joyful, obedient nature. I felt an instinctive love for God, and even wanted to become a minister when I grew up, although we had no religious training in our home.

"After the age of six, I became progressively more unhappy and troubled with terrifying nightmares, problems with my parents and even fears of being insane----a strange concern for such a young child.

"By the time I was an adolescent, I was severely emotionally disturbed, abused alcohol and drugs and needed psychiatric care. Although I ended the substance abuse and turned back to God and spiritual interests, I remained very emotionally unstable. For the next twenty-five years, I was plagued by violent mood swings, periods of black depression and outbursts of terrible rage. I remember telling a therapist that I sometimes felt as if a "terrible black monster" came out and took over my feelings and behavior. During those "spells," as I called them, I caused much harm to myself and others, physically and emotionally.

"Even in my better periods, I was prone to anxiety attacks, crippling fears and phobias and a chronic underlying anger. Longing to lead a spiritual life, I was nevertheless often rebellious and wild. There always seemed to be a destructive force pulling at me. I couldn't remain true to my own honorable values for long and did many things of which I was deeply ashamed.

"In 1987, I was hospitalized, diagnosed as having manic depressive disorder and put on medication which helped keep things at bay but never really cured me. I often "heard" thoughts in my mind which I knew were not auditory hallucinations. Sometimes when I was praying a thought would cut in saying, "Screw God" or something equally shocking. I had the peculiar sensation that I wasn't "alone" somehow in my mind and considered the possibility that I also suffered from multiple personalities. At times I felt an indescribably evil presence threatening to overcome and destroy me. All this sounded like a mental illness, but by this time I was a psychology major in college and I knew I really wasn't psychotic in spite of all these symptoms.

"When I finally read about spirit possession, I felt a deep inner sensing that at last I knew what was really wrong with me. But not knowing where to turn for help, I pushed the knowledge away as too fantastic to believe----until the blessed day I came in contact with people who were doing spirit releasement therapy. With them, I could finally face what I knew in my heart was true and discuss my problems openly without being thought insane. I was then guided to Dr. Hickman and Becky.

"During my session with them. Becky, in trance, saw that I was completely full of "dark forces" headed by a powerful demon who claimed to have possessed me at the age of seven. (When I later asked why this had happened to me, Becky saw that I had been offered to him in some sort of ceremony. This seemed to substantiate memories my sister had had that we were abused in satanic rituals as children.)

"My session had a powerful emotional impact on me. I knew the spirit that spoke through Becky. I had lived with it most of my life, always thinking it was a part of my own personality---a side of myself I hated and feared. I could fully relate to the anger and fear it expressed and could clearly see how even some of my most basic attitudes---rebellion against authority and chronic irritability---had been generated by it. When it claimed to have been sent to control and ultimately destroy me, I became vividly aware that this was the destructive force I had so often felt tearing at me. I felt a tremendous emotional and spiritual upheaval as this being at first defied Dr. Hickman and then eventually succumbed to her commands to leave and go to "The Light" As it departed, I felt an actual physical relief as if something was being wrenched from inside me. When it was over, I felt almost empty at first. Then an incredible Light filled my being. When I looked in the mirror, I was positively radiant. I felt physically light and emotionally free---an indescribable sensation.

"Since the depossession, I have been completely free of all the old symptoms for the first time in forty-one years. I have stopped taking any medication and am calm, stable and fully in control of my emotions. I feel at last like a mature version of that happy, loving child I remember being, free to love God and work for the good of humanity. I am considering actually becoming a minister as I dreamed of doing. It is the greatest blessing I can conceive of giving in my life."

Another note from Penny dated Nov. 29, 1993 told us that she continues to thrive without the old troublesome feeling.

This report was sent to me from Williamsport West Virginia. Terry Ash and Rhonda Newhart did the work for the daughter of a friend. Brittany, age six had been having night terrors for a long time. She had four to six episodes each and every night. She never remembered any of them. However she was afraid of monsters and it was a battle to get her to go to bed.

The remote depossession was done on a Wednesday afternoon. On Thursday, the mother reported that Brittany had had a milder form of night terror Wednesday night. The mother was advised to do a meditation putting Light around Brittany for several days. Terry also did several Light meditations for Brittany.

Two weeks later the mother called to ask whether another depossession had been done as Brittany had not had any further night terrors except the first night after Terry and Rhonda had done the depossession. From Terry's letter: "She just thanked us over and over again. Needless to say, Rhonda and I are just thrilled. Brittany is no longer afraid of monsters and goes to bed without a hassel. She's calmer, happier and sleeps well."

It is difficult to know how much effect was produced by the remote depossession and how much by the Light meditations. Additional research is needed in cases of this kind to help us determine the effectiveness of either or both of the therapies.

Charlene Ackerman, of Janesville, Wisconsin sent me this interesting report. It concerns Philip, age five, who had never spoken. He had made gutteral sounds that resembled words, and whined from time to time. His parents felt that he had the ability to speak, but simply was not able to get the words out.

Philip had been checked by a local child psychiatrist. He was thought not to be autistic and that he needed more stimulation and encouragement. His parents were eager to do whatever they could to help him. After discussion with the mother it was decided to do a remote scanning of Philip using his mother as the instrument.

The first entity contacted had been attached to his throat, creating a tightness and heaviness. This entity gave the name of "avoid" but also said that she was an old woman who had died alone of a heart attack. She also had an inability to speak. She went into "The Light" with the help of her father and told Philip she was sorry for holding him back.

Further scanning of Philip located a "dark" one in his head that gave the name of Short Circuit. This one left rather easily with the Forces of St. Michael who came for him. Then Philip's mother was directed to visualize filling Philip's body with Light and letting the Light expand to form a bubble of Light around him.

Two days after this session, Philip and his sister, Amy, were in the back seat of the car as his mother was waiting at school for another child. Amy was singing, and all of a sudden, Philip started singing. The first words he had ever spoken were from *Annie.* He sang *The Sun Will Come Out Tomorrow* and continued to sing the rest of the song. Philip still has difficulty talking, but sings a number of songs clearly. A second session of remote scanning of Philip did not produce further improvement. Perhaps at a later time there could be greater success, or the remaining problem may be from an entirely different source rather than from an attached entity.

This next case is from Olga Ivsin, M.D. of Chicago. She has given it the title of *Chinese Family Reunion.* The first request came from a Chinese man who could only communicate through an interpreter. This man is a Chinese herbalist, and horoscope reader, and explained that he thought his son, aged twenty-three, had a "bad Spirit" with him. He explained that his son, a graduate from an American College, and the owner manager of his own store, had trouble controlling his temper and had become aggressive toward his parents with whom he lives. The son had refused to come because he did not believe in any "spirits."

The father said that in Tiawan when a man is possessed by "bad spirits," one can find help in the Buddhist Temple where Masters know how to exorcise them telepathically, but that kind of service for Buddhists is difficult to find in America.

Dr. Ivsin, using her assistant as an instrument, through remote scanning located and removed three Earthbound entities from the son and helped them go on into "The Light."

Then the father requested that he too be scanned. Four entities were released from the father. The following week the Chinese mother through an interpreter requested a remote procedure for herself. Two Earthbound entities were released to "The Light" from the mother.

Two weeks later the son, who spoke fluent English, requested a remote depossession for himself. One more Earthbound and one demonic were found and released. A month later the father reported a cooperative and helpful son, and that the family was closer than it had ever been.

In November of 1993, I and ten other hypnotherapists were together during a Caribbean cruise. In addition to visiting the islands, we had several training and practice hypnotherapy sessions.

On November 13, while we were at sea between Martinique and Puerto Rico, one of our group, Ivan D. Goochey, of Bennet, Nebraska, stated that he would like to be the hypnotized scanner to see whether he could do this work. This was a first time for him.

He selected a friend living in Wisconsin, a forty-one year old female, who was on medical leave as a pilot from United Airlines. She had developed severe sinus and other respiratory problems after being exposed to the heavy smoking of the pilot with whom she flew.

Under hypnosis, Ivan easily located the essence of the target subject, but required more time to locate the physical body. Once this was accomplished, an entity was found in the lower throat area. In dialoguing with this entity we were told that he had once had his own body, named Maurice, that he was a push-cart peddler on the streets of a French city. He was run down and killed by a large vehicle of some sort. He mentioned viewing the remains of his physical body scattered about the street and concluded that he could never return to that body. A sympathetic lady, D...., standing on the street witnessed the accident. Maurice "joined" her via the throat chakra. Maurice was willing to look for others and noted a reluctant one in the pelvic area. He then was willing to move to "The Light" to join loved ones who came for him.

This entity located in the pelvic area said her name was Neandra. She stated (speaking through Ivan) that she had been with D.... since Ancient Egyptian times. She had been a hand maiden to an Egyptian lady of some social and political standing. During a journey, she had been raped and beaten to death by male attendants to the same lady. She felt angry and betrayed. At once, at her death, Neandra entered the body of her mistress via the lower chakra. Here she caused pain and discomfort in an effort to get revenge because of her mistress' indifference during the rape attack.

Neandra was very reluctant to leave, but did agree after a time to leave and go to "The Light" with an angelic being who greeted her.

Attention was then moved to the sinus/eye area. Here there appeared to be a swarm of tiny entities resembling a swarm of bees, none of which had ever had a human body of their own. We were able to dialogue with a spokesperson for the swarm. Extreme resistance was finally overcome with the help of St. Michael's Forces. The swarm left with their spokesman leader. We felt that these may have been thought forms.

After these intruders were gone, we asked the healing angels to bring healing Light and fill the spaces vacated by those who had departed.

Four days after returning home, Ivan was visited by D... No details were given to D.... except that he had participated in a healing for her without mentioning the date. He inquired whether she had experienced anything unusual. Her response follows.

"On Saturday, Nov. 13, I had not felt well and remained abed. At about 10:30 (the time we were doing the remote depossession) I broke out in a sweat which persisted until noon-time. Quantities of mucus were purged from my throat and my voice became "froggy" for a while. I experienced some mild abdominal pain, then, within 24 hours, my menstrual cycle began. Usually I had been experiencing bloating, cramps and clotting. This period was atypical in that it was pain free with no clotting or bloating

"On Sunday, (the day following the depossession) my sinus drained and felt clear. Another thing, I have had a burst of sustained energy---completely atypical for me."

This interesting case came from Dr. Brian G. Knight of Corpus Christi Texas. One of his patients, a boy of eight, had complained ot abdominal pain for six months. Examination could find no organic cause for his complaint. Spirit releasement was discussed with his mother. She agreed to be the one who scanned her son.

She found a small red area in her son's upper abdomen. It was difficult to obtain any response to questions. (Dr. Knight commented to me that the questions were probably too forceful.) Eventually a small voice did come out sounding very uncertain and frightened. The entity identified itself as Mickie. He had been struck by a truck when he was five in Atlanta, Georgia. He did not know how he got from Georgia to Texas, but answered, "I only wanted to be with 'nother little boy."

Mickie had no knowledge of what to expect at death, but when someone from "the Light" was called for and Mickie was told to look up, he cried out, "Gramps" happily and joined Gramps. (The boys pain left and did not recur.)

Linda Cox, of Oxford, Michigan contributed three cases. She included a note also of how she uses a protective ritual. Even though I do not use such, it is quite possible that others would wish to do so.

"I advise everyone to work with the Light of God stating, for example, 'I am of Light/God and only Light will enter. I serve only the Light/God. The Light of God always prevails.' I also use the following prayer and give a copy to my clients."

"The Light of God surrounds me. The love of God enfolds me.
The power of God protects me. The presence of God watches
over me. Where ever I am, God is."

Linda adds: "Remote Spirit Releasement is a wondereful healing modality and very easily learned. Thank you, Irene, for once more being a catalyst so that others may learn and have the opportunity to be free in body, mind and spirit. Spirit Releasement Therapy is indeed the most important work we have to do in our healing. We must release any interference in order to get to the cause of any condition. I see it as a mission of love and one of the most rewarding modalities we can be a part of. God Bless You."

The first case from Linda was a mother requesting help for her two teenage daughters and for herself. She said that her husband had committed suicide and that the energy in the home had been very "negative" since. She also remarked that since a seance that had been conducted in their home recently, the girls were fearful and suffering nightmares.

Linda began with the invocation of Light for protection and assistance. She asked the Warrier Angels to stand against the darkness together with all in "The Light" who loved and supported the mother and her daughters to guide the healing.

The mother was cleared first, releasing the husband who was stuck in the trauma of his suicide. He also had a number of demonic beings attached to him.

The mother was the instrument for remote scanning of the daughters. Many entities were found with the daughters and released to "The Light." They were then filled with Light and wrapped with spiritual bandages. It was further found that a number of lost souls had come to the home during the seance, had decided they liked it and had stayed. These were released to "The Light." (The person who had conducted the seance had not invoked the Light or made any effort to protect those present. During the seance, a white candle that had been burning, suddenly started to smoke and turned completely black. This frightened everyone. The seance ended in fear. Their fear became a doorway for lost souls and "dark" ones to enter.)

Linda then proceeded to clear all inappropriate energy from the house, the pets, the plants, the car etc. A few days later, the mother reported that the daughters were no longer fearful or having nightmares and that the house was peaceful and enjoyable.

The second case from Linda was a ten-year-old boy who was out of control. He was described as wild, having fits, kicking the walls and was a major problem at school. The mother stated that he had been diagnosed as (ADD) and Ritalin had been recommended. She was reluctant to start him on Ritalin and felt that he was possessed by something evil.

The session was started with an invocation of the Light for protection and assistance. The Warrier Angels were called to stand against the darkness. We called also for all in "The Light" who loved and supported her son to guide the healing. Permission was requested of the Boy's high self to scan him. Permission was granted.

The mother scanned her son and found him completely filled with darkness tinged with red. A dearly beloved grandmother who had committed suicide was found to be with the boy. His anger at her leaving together with her anger and the trauma of her death had allowed major darkness to enter. Grandmother and the other energies were released to "The Light." The healing angels were asked to fill the son with healing Light wherever it was needed. The mother was further instructed to visualize wrappings for her son, both inside and out, of spiritual bandages----purple for forgiveness----pink for unconditional love----emerald green for healing at all levels----turquoise for change, integration and processing of the work that had been done.

Inappropriate energy was cleared from their home, car, animals, plants and even his school. The mother was further advised to keep herself and the boy filled with Light and continue to use the spiritual bandages.

When the mother picked up her son at school on the day of lthe remote depossession, she learned that he had slept during the time of the procedure (most unusual for him) and was calm and loving. As a result of the depossession, he no longer needed to be given Ritalin.

Linda added that she had had other ADD children as clients who after spirit releasement had become able to concentrate and that both their behavior and their grades had markedly improved.

Linda's third case report concerned a thirty-year-old man who was diagnosed as schizophrenic. Since his grandfather's death eleven years earlier, he had displayed the old man's characteristics and mannerisms. Grandfather had been senile, full of anger and a heavy smoker. The grandson began to smoke right after losing his grandfather and to exhibit nearly identical ways of sitting, talking and expressing anger.

The young man's sister was the instrument for the remote. Protection was invoked, permission to scan was asked and granted. Grandfather was there and was very reluctant to leave. He would say, "I'm not going anywhere. I like it here." When he recognized his wife who came for him in "The Light," he began to cry and was willing to leave with her. Once he was in "The Light." he was completely healed in mind and body and said he was sorry for staying with his grandson and was pleased to be able to leave.

A few weeks later, the family reported that though there were still some remaining emotional problems, the grandfather was gone and that they had had the best family Thanksgiving ever.

Henry Smith-Rohrberg of West Falmouth, Massachusetts, has developed a novel method of remote depossession. Some, he tells me, he has done through meditation, some over the telephone and others while he and his wife are jogging.

One young girl developed behavior suggesting possession by "dark" spirits after a skiing accident. Her injuries included brain concussion, fractured vertebra and shoulder nerve damage. After the accident several things changed. Her personality deteriorated, she was depressed and physically ill a great part of the time. Her grades declined and she developed a dark tendency to talk about the Void, death, murder, suicide and blackness. She became progressively alienated from everyone except a few rather like-minded peers.

Henry and his wife, Karen, did several remote releasements involving this girl while they were running. The entities found and released included demonic entities, Earthbound entities, past life entities and soul fragment entities. As a result of several 'Road-Running Remote Releasements," the change in the troubled teen-ager were described as miraculous.

To quote from Henry's letter: "Since beginning my Spirit Releasement work as a clinical psychologist, I have been very busy and often have requests outside my regular office work. I have done releasements on the telephone in the middle of the night, both direct and remote. I have worked through my lunch hour, and sometimes I have been so tired that I have fallen asleep for seconds at a time while facilitating releasements. No matter, the techniques still work and I feel blessed and privileged to be able to do such effective and redemptive playwork for Soul's advancement and ascension.

"The Spirit Releasement techniques that I and others are using are evolving quickly and wonderfully. This work is truly safe and easy. Irene Hickman recently asked me: "Can you posssibly imagine a more exciting work than what we are doing with releasements?" My answer is that I cannot."

Chapter Ten

IS REMOTE DEPOSSESSION MEDDLING?

At first thought, it would appear that scanning for entities without the conscious knowledge of the person being scanned is indeed meddling. I am of the opinion that if the Ten Commandments were to be expanded to twelve of fifteen, one of the added ones would be "Thou shalt not meddle."

In the same way, any distant healing or even prayer for another who has not themselves asked for it could be called meddling. I feel that special effort should be made to avoid this kind of intrusion into the psyche of another.

My special effort is to always ask permission of the high self of one to be scanned. This high self has the best interests of the whole person as a first concern and would not give permission if harm would result. Whenever we receive a response of "No" or "Not now" when permission is asked, we do not proceed.

As long as the attached entities are there as uninvited trespassing intruders, it seems acceptable to locate them and sell them on the idea of going to a better place. Their departure is a benefit both to the host and to the entity who was attached.

So many times the person with the attached entity does not suspect its presence and does not think of seeking help through depossession. It seems to me that it is only necessary to avoid acting from our personal ego stance of "Look! See what I can do!" Instead we need to be concerned that all beings touched in any way are being served lovingly.

DOES DEPOSSESSION ALWAYS WORK SUCCESSFULLY?

A common axiom---"Never say never, never say always" probably applies here. It would be gratifying to receive 100% feedback reports of obvious beneficial results. I do not know how to accomplish this.

Many depossession sessions in which I have been involved have resulted in no feedback at all. Of the number I have received, only two appeared not to be successful. During one of my out-of-town workshops, we did a remote depossession on the two sons of a participant in the workshop. She phoned me the next day saying that one son showed marked change for the better, but that the second son was about the same. Our conversation continued during which she revealed that the unimproved son had been drinking after the time of our session. We could conclude that we were unsuccessful in removing all trespassing entities or that his drinking had opened him to further entity intrusion.

Another not-so-glowing report followed a session we did in Chicago at the MidAmerica Hypnosis Conference in August, 1993. Corinne S. of Ohio, requested a remote for her brother, Stephen.

Stephen had a long history of serious emotional disturbance. He had had thirteen electric convulsive treatments in an attempt to stabilize him. On the day we did the remote releasement he stopped drinking and remained free of drink for six days. This was unusual for him as he was in the habit of drinking regularly. Then he began to drink again, his irrational behavior escalated and he required hospitalization.

Some excerpts from Corinne's letter: "He often tried to make me angry. When my latest response wasn't anger, but the expectation that he would be held responsible for whatever he said and did and that I would not bail him out, his anger escalated. The same or different spirits seemed to have returned in full force. I was able to tell him to leave.

"I was reminded of one of the statements from your talk in Chicago---that one should always question everything. I questioned whether spirit releasements done for my brother would really help him as they did not make him "well." As I write this I recall something else that you said that made an impact on me, that many times we are stubborn and had to face the lessons over and over, "God's will, not mine, be done." I have come to believe that there is only so much that you can do for another person, no matter how much you love them. My brother did arouse anger in me and I have been struggling with that and have finally found peace in knowing I have done all I could do for him. I continue to send him Love and Light but have learned that now is the time to let go. If a person that you are doing remotes on doesn't want to change and grow, then aren't you only temporarily clearing him of spirits? His will and refusal to change starts the cycle all over again, doesn't it? Even if that is so, real and profound changes have occurred within my family.

"Stephen is now faced with a family whose buttons have been disconnected......This has been a difficult letter to write. I found it very enlightening for me. My original intent was to question whether doing remotes was something that falls within human will or God's will, as the results were not what I had hoped for. I now have renewed faith in the larger plan and note the very significant changes that have helped the whole family deal with Stephen in a new way."

These two reports are balanced by many favorable ones. Most of my life I have felt that I would like to change the world for the better. I now feel that I have found the right way to do it.

Depossession and remote depossession as described here is not new. Carl Wickland, M.D., a psychiatrist with Spiritualist connections, worked with many severely disturbed patients whom he thought to be possessed. For thirty years, his wife, Anna, a trance medium, would allow spirits possessing another person to speak through her voice. Dr. Wickland would dialogue with these spirits, educate them about a better place and persuade them to leave their host.

His book, *Thirty Years Among the Dead, (1924)*, describes in detail his dialogues with spirits and his theory of the role of spirit possession in mental illness.

Between 1900 and 1945, Edgar Cayce, an American mystic, gave over fourteen thousand trance "readings" mostly for diagnosing and recommending treatment for sick people who had not been helped by traditional medicine. Cayce was a Bible expert and Southern Fundamentalist Christian when awake. In deep hypnosis he had much to say indicating that our illnesses could be caused by past-life trauma or possessing entities.

My own first acquaintance with the spirit possession concept was from hearing a lecture by Dr. Edith Fiore at the First International Congress on Alternative Therapies in Sao Paulo, Brazil in 1985. Dr. Fiore told of having done several hundred depossessions and said, "But I've never met a demon."

In July of 1988, I heard her speak on this subject again at the First International Conference on Paranormal Research in Fort Collins, Colorado. Here she stated that she had dealt with a few "demons."

At that conference, several other speakers discussed possession cases. These included Dr. Hugh Harmon, Dr. Louise Ireland-Frey, Dr. Bill Baldwin and myself. My case was one in which an aborted foetus had claimed to have entered her sister, six years younger, and had caused various problems. I could only describe this case. I had had no training or understanding as to how to deal with the problem.

I was particularly impressed with the presentations of Dr. Bill Baldwin, both during a lecture and a two-hour workshop. In May of the following year, I had training with Bill.

I realize that my work with remote depossession more closely resembles that of Dr. Wickland than the others, but differs in that Dr. Wickland would use a mild electric shock to drive out recalcitrant spirits, who would then enter Anna. He would then remove them from Anna sending them to a better place.

Dr. Fiore was the first to give us good useable instructions as to how to deal with possessing entities in her book, *The Unquiet Dead, (1987.)* She was Baldwin's mentor. He in turn has developed his own style and methods which he relates in his book, *Spirit Releasement Therapy, (1993).* I in turn have modified what I have learned. Those I have taught each develop their own therapy style and method. Who is to say which style is best? Perhaps we each must find that which works best for us.

Chapter Eleven

DISCUSSION

As I think about the growing interest in alternative therapies, in hypnotherapy in particular and in spirit releasement or depossession, I am led to wonder about the possible effects of this book. Will it fade away and be barely remembered by a few or will it unleash a deluge of requests for remote depossessions far greater than the few trained therapists can accept?

One possible happening is that some people will seek depossession therapy when the problem has a different origin---a different cause.

Possession by an entity, Earthbound or demonic, is not the only cause of our disturbed health or our undesirable behavior. It is likely to be tempting for many people to seek to put any blame for unwise choices or behaviour onto an entity.

My observation is that some people can completely disrupt their lives from their deliberate choices without needing any extraneous motivation. For these cases, the pattern of reward--punishment popular with Behaviorist therapists could be effective.

For some, the origin of problems is to be found in traumatic events in the present life---possibly as far back as early childhood, infancy, birth or intra-uterine. Dr. Fiore describes a case whose low self-esteem and feeling of unworthiness of being loved originated when an attempt was made to abort her.

Many problems have their origin in emotional traumas in earlier lives. A masterful two-volume set of books, *Regression Therapy, A Handbook for Professionals, (1993)* by Winifred B.

Lucas, Ph.D., is a splendid sourcebook for those who would like to understand regression into former lives.

Unless we realize that there are several possible causes for problems, we may be doing our patients or clients a disservice. Entity releasement will not solve a problem unless that problem has been caused by the presence of an attached entity. It is essential that we find and deal with the cause, whatever it is.

I expect the interest in depossession to continue to grow and attract an ever increasing number of therapists. If this happens we can expect to produce ever greater and greater changes in the lives of the world's population.

WHAT NEXT?

A most encouraging development is the establishment in 1992 of the National Institute of Health, Office of Alternative Medicine, with Larry Dossey, M.D. as its co-chairman. Dr. Dossey, author of *Beyond Illness, (1982),* now living in Santa Fe, N.M. has long recognized the shortcoming of medical practice that fails to give proper recognition of the role of mind and emotions in human health problems.

Iowa Senator, Tom Harkin, chairman of the Senate Appropriations Committee which funds NIH, favors this investigation of alternative therapies. His committee issued this mandate.

"The committee is not satisfied that the conventional medical community as symbolized by NIH has fully explored the potential of unconventional medical practices....In order to more adequately explore these unconventional medical practices, the committee requests that the NIH establish within the office of the director an office to fully investigate and validate these practices. The committee further directs that the NIH convene and establish an advisory program to fully test the most promising unconventional medical practices."

Panels within the Office of Alternative Medicine are now studying various alternatives. Their report is expected to be completed and available to the public by late Spring 1994.

With the growing number of Americans who are dissatisfied, even outraged at the ever rising cost, the side-effects, the impersonal treatment and frequent ineffectiveness of mainstream American medicine, it surely is time to examine alternative options.

Further research will surely find that some methods now thought of as "fringe" will be not only effective, but also safe and less expensive. The Office of Alternative Medicine appears to be heading toward change in medical treatment or in helping to determine what the new changes will be.

Our depossession work has some support from the writing of Dr. Ed Murphy, Associate Professor of Bible and Missions at San Jose Christian College. His scholarly book, *The Handbook for Spiritual Warfare, (1992),* includes many cases in which what he calls "demons" affect their hosts unfavorably.

Many that he calls "demons" appear to me to be similar to the Earthbound entities we encountner in our depossession work. Dr. Murphy's belief system appears to prevent him from considering that some attachments are Earthbound or "Unclean Spirits." He seems to consider all evidence of the presence of attached beings as produced by "demons."

Dr. Murphy includes many interesting demon names such as: Anger, Anxiety Bitterness, Confusion, Deception, Dread, Fear, Liar, Lust, Panic, Rejection, Terror, Worry and others. We occasionally encounter entities having similar names. We have an entirely different method of freeing them from their human hosts.

Dr. Murphy describes prayers and rituals used for "deliverance" of afflicted ones without including that there is a place for the entities to go after leaving their human hosts. I feel that Dr. Murphy's greatest contribution to the effort we are making is his stressing that attachments are real, even though he seems to have failed to recognize that entities come in two (sometimes more) species---the Earthbound and the "dark" or demonic and that these differing beings need to be treated differently.

As more and more therapists become trained in remote depossession, the possible changes that can be produced are infinite in number.

TRANSFORMATION

Since we have seen results from remote depossession transforming insane criminals, critically ill patients, incorrigable children, violent spouses, pain-racked clients, depressed and discouraged folks and others into productive, healthy harmonious persons, I would expect that the possibility for widespread beneficial changes is limited only by the number of people able and willing to do this work.

It appears that worldwide transformation has become possible through the use of the techniques of remote depossession.

Chapter Twelve

DEPOSSESSING A "DEMON"

All of the "dark" entities we encounter could be classified as of the "demon" species. Very few have exhibited the characteristics usually attributed to demons.

However, in the nearly five years I have been working with depossession techniques, I have encountered four who might justly deserve the title. Three were found during one-on-one spirit releasements. The fourth appeared during a "remote."

Lee, who had been doing releasements, told us that he had felt one entity he had been releasing from another person enter him. We agreed to check him by remote techniques while he was in the room with us.

Becky had done many "remotes." She was hypnotized and her "Inner Wisdom" was consulted as to whether she should scan Lee. Approval was given. Permission was asked of Lee's high self and this was granted. Becky began scanning Lee.

Becky I see a real darkness that instead of being inside, it seems to be coming around from the back.

Dr. H. Anything else?

Becky Something about his hips.

Dr. H. Any place else?

Becky No.

Dr. H. I call now to you who is causing this darkness that appears to come around from the back-----(At this point, Becky began moaning, flinging her arms about and appeared to be distressed.) Have you ever had your own human body? (More moaning and rolling from side to side.) I call now on the Forces of St. Michael, the mighty rescue angels, to come and surround this one with a net of Light. (More outcries as of pain.) STOP your moaning.

(Becky continued expressions of discomfort.)

Hush it! (outcries containing a note of fear) I know you're frightened, but you're not hurting, so hush it! I ask the Forces of St. Michael to pull the net tighter and tighter. Answer me. Have you ever had your own human physical body?

(More outcries of pain and then crying.)

Hush it! Calm down. I want to communicate with you and I want you to answer my questions. Enough of this. I know you're frightened, but you have nothing to be frightened about. All we have done is put a net of Light around you just to persuade you to answer the questions promptly and truthfully. As soon as you are willing to do that you will be quite comfortable. What is your purpose here in Lee?

Entity (shouting) Whooohaw.

Dr. H. You're trying to frighten me aren't you? Calm down. Answer my question. What is your purpose here in Lee?

Entity (Shouting) Kill him!

Dr. H. You tried, didn't you?

Entity Yes.

Dr. H. Did you succeed?

Entity Not yet.

Dr. H. Was that your assignment?

Entity I will succeed. I always succeed.

Dr. H. Who gave you that assignment?

Entity Rachaah.

Dr. H What happens when you give the name of your boss?

Entity Nothing to you.

Dr. H. Does the boss like you to give out his name?

Entity He doesn't care. (Threatening tone.) You can't hurt us.

Dr. H. I'm not trying to hurt you. Will you answer my questions calmly now?

Entity Yes.

Dr. H. How old was Lee when you joined him?

Entity Forty.

Dr. H. Was it your decision to kill him or was it an order from your boss?

Entity I hate him.

Dr. H. What reason do you have for hating him?

Entity His work.

Dr. H. What is it about his work that you hate? (More loud long outcries) Stop your moaning. Answer my question. What is it about his work that you hate?

Entity (Growling) You know.

Dr. H. I'm asking and I expect an answer.

Entity (Shouting) You too!

Dr. H. You don't like what I do? What do you know about what I do?

Entity (Still shouting) I know.

Dr. H. What do you observe about Lee's work and mine?

Entity (Angry shouting) You mess up our plans. You ruin things.

Dr. H. I would like you to look a little more carefully at the result.

Entity (Unintelligible babbling)

Dr. H. What really happens because of Lee's work and mine?

Entity (Loud frightened tone) I'm not going to talk to you.

Dr. H. You will talk to me. The net around you gets tighter and tighter until you decide to answer calmly without all this moaning and screaming. I want you to admit what we really do.

Entity (Frightened) I don't like this.

Dr. H. You're getting scared, aren't you? You thought you could frighten me and Lee, but it didn't work. I suspect that you're really scared of your boss. He might be watching. What will your boss do if you don't follow orders?

Entity (Shouting) I follow orders.

Dr. H. What would happen if you didn't (Low growling moans) Answer me and answer me truthfully.

Entity No!

Dr. H. No, what?

Entity I won't answer you.

Dr. H. The net grows tighter and tighter until you do. (Loud moaning as in pain) Stop the moaning. (More painful crying) Now you really know the kind of work that Lee and I do.

Entity I hate it! I hate it! I hate it!

Dr. H. But you're just afraid that you're not going to succeed. You're afraid that you might be punished if you don't succeed. Isn't that right?

Entity (Loud snort.)

Dr. H. Answer me. Feel the fear---much more than your anger and hate. Feel that fear---feel it. (Painful outcries) Stop your moaning. I can help you.

Entity You can't help me.

Dr. H. Hush your moaning and listen. I have helped others and Lee has helped others. (Sobbing) You have a horrible existence. You take orders from horrible beings. You have failed and are full of fear. Right? You're failing at your job.

Entity (Tone of desperation) It's not my fault. It's not my fault.

Dr. H. I didn't ask you whose fault it was. You are failing, right? (Heavy gasping) Aren't you failing?

Entity I just have to get back. I just have to get back.

Dr. H. Back where?

Entity In him.

Dr. H. But you are failing and you're scared. Feel the fear. I can help you be relieved of this fear.

Entity How can you help me?

Dr. H. Just try it. Listen to me and follow a few simple suggestions. Haven't you had enough of this fear?

Entity (Sobbing) Yessss.

Dr. H. And you know what will happen to you if you fail.

Entity Yes.

Dr. H. There is another option. You don't have to continue to fear that awful place where beings like you are sent for punishment.

Entity There is nothing else. There is nothing else. (Still frantic with fear.)

Dr. H. You have believed a lie if you have believed that What were you told about the Light?

Entity NO! Stay away!

Dr. H. What reason were you given for staying away?

Entity It melts you.

Dr. H. I does not melt you. That's a lie. I ask the Forces of St. Michael–– (Interruptions of loud NO, NO, NO.) Stop your moaning and pay attention to what's happening---- to send a shaft of Light right through you, and another and another. (More loud outcries) Hush up and notice. Did it melt you?

Entity (Quieter now) Nooo.

Dr. H. You have been believing a lie. That was a deception. You have allowed yourself to be deceived. I can help you escape from the liars and deceivers who give you orders. You were told that the Light would destroy you and cause you to melt, but that didn't happen. Do you recognize that you have been deceived? Do you recognize the lie.

Entity Yes.

Dr. H. Those beings around you that have you inclosed in the net and have shown you the shafts of Light are here to rescue you from your horrible and fearful existence.

Entity It's awful.

Dr. H. Of course it's awful, but you don't need to continue. You've had enough haven't you?

Entity (Sadly) Yes.

Dr. H. Become aware of the beings around you. Communicate with them. Ask them why they are here. Ask them what they can do for you. What do they say?

Entity (Softly) They say they can help me.

Dr. H. Ask them how they can help you.

Entity How do I know they are right?

Dr. H. You have found out that what you have been believing was not the truth. Take a chance. Ask them about their existence---whether they are filled with fear. What do they tell you?

Entity They say it's better and that I can go with them.

Dr. H. As you get ready to go with them, look deep within the very center of your being and tell me what you find there.

Entity Nothing.

Dr. H. Look deeper. Now what do you find?

Entity A little spark.

Dr. H. You were told that Light would melt you and here it is within you. It has been there all the time. You have allowed the deceivers to convince you of a lie. That was a mistake. You're not all black. Watch that spark that you have found at your center and tell me what changes take place.

Entity It's growing.

Dr. H. What else? Is it uncomfortable in any way?

Entity No.

Dr. H. What's happening to your darkness?

Entity It's going.

Dr. H. Tell me when it's all gone. What's happening?

Entity It's gone.

Dr. H. Look at the beings around you now and at yourself. Now do you believe that you might belong with them?

Entity Yes.

Dr. H. Now, before you leave Lee, look around inside his body and around the periphery. Are there any others of your former dark nature in or around Lee?

Entity No.

Dr. H. Prepare to leave and as you leave Lee's body, call out through the universe for all beings of your former dark nature to come and join you and go home with the rescuers. Are you calling?

Entity Yes.

Dr. H. Are they coming?

Entity Yes.

Dr. H. How many?

Entity Too many to count.

Dr. H. Remind each one of them to look deep inside and find their true nature at their core. Go in peace with our love and blessing. I now ask the healing angels to fill the space in Lee vacated by the fearful one.

Now Becky, locate the brightest spot of Lee's spirit, focus on it and let it expand filling his body and extending out around his body at least arm's length in all directions.

Now fill your own body in the same way and tell me when it is finished. Prepare to waken. The forces within your body will equalize as you waken. In the future, if we have an occasion to work together, I will take your left hand in my left hand and count from one to twenty-one. If you are willing, by the time I reach twenty-one you will be deeply relaxed in hypnosis and ready for the work. If for any reason you are not willing, then neither I nor anyone else will be able to induce this state in you. You will always remain in control, free to accept or reject whatever I or anyone else says to you, free to answer questions or to remain silent. Waken now as I count backward from twenty-one to one using alternate figures. (Backward count)

The next morning, Lee reported that he felt completely free of the entity's presence and influence. We have had no later report. During the session, Becky had appeared to be in great distress, but as soon as the depossession was finished, all of those characteristics and behavior were completely gone. There has been absolutely no residual undesireable effects of this experience.

At times it seems essential to be very firm with the "dark" entities. It is helpful to know that they operate almost entirely on fear, and that we can use that fear to convince them that we have a better offer for them. No matter how assertive I might have been initially, I find that I can feel true compassion for them as they begin to break free.

Chapter Thirteen

MORE REMOTE DEPOSSESSIONS

We are sometimes asked to scan a house or other building for the presence of entities. Emma, a friend, consented to do this even though she had never done it before. After inducing hypnosis, her Inner Wisdom was consulted, and the project was approved.

Emma sent a bit of her consciousness to the address given and found evidence of an Earthbound, George, who said he had died there fifteen years earlier and was not interested in a better place. He admitted that he had heard aboout "The Light" but that he didn't pay attention then and didn't want to hear about it now.

I explained that this was no longer his house, to which he replied that he didn't want to leave it---even to go to a better place. I called for anyone in "The Light" who cared for George to come for him. Alice came.

Geo. She looks happy, but I still don't want to go.

Dr. H. I will make you an offer, George. If you don't like it there after you go for a visit, you can come back again. (He accepted the offer.) George, look around the house and outside for any others you can take with you.

Geo. There's an infant outside, near a body of water, who drowned. He will go with me.

Dr. H. Continue to look around the house for any others. Are there others?

Geo. Still looking. There's a shadow in the attic that never had a body.

Dr. H. Go in peace with our love and blessing. Thank you for taking the infant with you. I now ask the healing angels to fill the house with Light.

We then asked Emma to explore the attic. There we found one that never had a body and had no name and didn't want to talk. St. Michael's Forces were called to bring the net of Light and tighten it if needed to get this one to answer questions.

Dr. H. What is your purpose here in this house?

Entity Who are you?

Dr. H. I am here to help you.

Entity This is my place.

Dr. H. Who sent you to this place?

Entity The "dark" one.

Dr. H. What is your purpose---your assignment?

Entity To make the people sick.

Dr. H. What were you told about the Light?

Entity It doesn't exist. (I asked the Forces of St. Michael to send several shafts of light through this entity.)

Dr. H. Does "Light" exist?

Entity I suppose it does.

Dr. H. You have been lied to. You have believed a lie. How do you like that?

Entity I'm confused.

Dr. H. Do you like being deceived?

Entity No.

Dr. H. What were you told about staying hidden and remaining secret?

Entity I don't want to think about it. I could be punished. I haven't been there but I know about it.

Dr. H. Now that you have been found, you might have to go there. Would you like that?

Entity No.

Dr. H. Fortunately, you have another option. You don't have to go to that place of punishment. There is another choice. You can go with those who have come for you. Become aware of the beings who have brought the net and surrounded you. How do they look to you?

Entity There is Light coming from their eyes.

Dr. H. Look at yourself. How do you look?

Entity I'm dark.

Dr. H. Would you like to look like them?

Entity I would like to try it.

Dr. H. Look deep within yourself---to your very center and tell me what you find.

Entity There's a very small spot of Light.

Dr. H. Keep watching that small spot of Light and tell me what happens.

Entity It's expanding inside me.

Dr. H. What is happening to your blackness?

Entity It's almost gone.

Dr. H. Now look at yourself and at those outside. Are you more like them now?

Entity Oh yes.

Dr. H. Prepare to go with them, but first I would like you to look all around to see if there are others of your nature there with you.

Entity There are some others.

Dr. H. Call them to you and ask each one to look inside to their center as you did. Are they changing too? (The others all changed to bright beings also. Further searching found no other "dark" ones.) Go in peace with our love and blessing. As you leave this house, call out through the universe for all those of your former "dark" nature to come and go with you. Are you calling?

Entity Yes.

Dr. H. Are they coming?

Entity Yes.

Dr. H. How many?

Entity Over one hundred.

Dr. H. Tell them to look inside and find their true nature at their core. Now I ask the healing angels to bring Light and fill the spaces vacated by those beings that have departed.

Emma also agreed to scan Y..., who, when she had been hypnotized and regressed to age ten, had become very upset, crying out: "Help me! Help me! I'm so scared! I'm so scared! for several minutes. She then laid her head to one side on her hands saying, "And then I died." Y... was not able to give me much information about her reason for being so scared, but after several repetitions of the emotional outburst, she became more calm. After she was wakened from the hypnosis, she told me that her father had died of liver cancer when she was ten and that her own liver was beginning to hurt. She was concerned that she might be developing cancer. It was the next day that we did the remote scanning of Y... with Emma as the instrument for the remote. Emma's Inner Wisdom approved the effort. Y...'s high self granted permission.

Emma began scanning Y... and noticed a dark area in her liver. We asked for identification, which was difficult because of the Chinese language. We did learn that the entity in Y...'s liver was her father. We then asked that the entity just put the thoughts into Emma's head and let her speak for him. The name was difficult but sounded rather like Lo Yuin.

Dr. H. Go to the time when your body died.

Lo I didn't want to die.

Dr. H. What did happen when your body died?

Lo Very sick. Consumed with pain.

Dr. H. Go to the moment your body stopped functioning. What are you experiencing?

Lo I don't want to die. I don't want to die. I want to get back in this body.

Dr. H. Your body has died, but <u>you</u> haven't. Isn't that true? What happens after your body dies?

Lo I want to live. I want to find somebody. If I can live inside someone else, then I can continue to live.

Dr. H. Do you find someone else?

Lo I attach myself to my daughter.

Dr. H. There is a much better thing for you to do, father of Y... There is a place called "The Light" that has been designed and prepared especially for those of us whose bodies die. Do you know about "The Light?"

Lo Yes, but I want to live.

Dr. H. Do you realize that you are causing harm to your daughter?

Lo No.

Dr. H. She is having pain such as you had. Did you want to cause that?

Lo No.

Dr. H. Are you willing to stop causing the pain?

Lo Yes.

Dr. H. In order to stop causing the pain, you must leave your daughter's body. Are you willing to do that?

Lo Where would I go?

Dr. H. There is a good place for you to go. It is called "The Light." I call now on anyone in "The Light" who cares about Lo Yuin to come for him, greet him and help him with his transition. Look up. Tell me what you see.

Lo There's a long line of my ancestors.

Dr. H. What do they tell you?

Lo They tell me to reach out to them and come.

Dr. H. Are you willing to do that?

Lo (Hesitating) Yes.

Dr. H. Ask them what their existence is like.

Lo They say it is joyful and peaceful.

Dr. H Do you think that might be better than staying with your daughter?

Lo I'll miss her.

Dr. H. Will it be better for you?

Lo It looks like it.

Dr. H. Then if you are willing to leave your daughter and go with your ancestors, your daughter will stop hurting, and you will have a good existence. Before you go I ask you to look around and inside Y...'s body. Tell me are there any others there that are not a part of

Y... ?

Lo There's something in her right ankle.

Dr. H. Is that one like you in nature or different.

Lo Different.

Dr. H. Are there any others of your nature?

Lo No.

Dr. H. Lo Yuin, take the hands of your ancestors and go in peace with our love and blessing. I ask the healing angels to bring healing Light and fill the space that was occupied by Lo Yuin.

I call now to you in Y...'s ankle. Come forward and identify yourself. Do you have a name?

Entity No. (This one was easily persuaded to go with St. Michael's Forces that we called, and seemed pleased to have the opportunity to be free of the dark commanders. When this one called out through the universe for other "dark" ones to come, first there were six who came and then twenty-five more.)

Repeated scanning disclosed another Earthbound in the back of Y...'s neck that told us that they had joined Y... when she was eight. When we called for someone in "the Light" to come for this one, it was Lo Yuin who came.

As usual, we asked that the spaces be filled with healing Light by the healing angels, and that Y...'s body be filled and surrounded with a capsule of Light.

After clearing Y... we asked Emma to fill her own body with Light as she had filled Y... and the house she had scanned. Emma found a shadow in her own head. This was Joe, who had died falling down a stairs and had joined Emma whan she was twenty-three. Joe's wife came for him from "The Light" and he left with her. Light then filled Emma's body completely and extended beyond to inclose her in a bubble or capsule of Light.

She was then given the usual post-hypnotic suggestions about future work together and wakened. Y... reported that she had been having some discomfort in her ankle and back of her neck as well as her liver. After this session, she was completely pain free---warm and comfortable.

I am including these next cases because they contain some unusual elements. Dee, (not her real name) agreed somewhat reluctantly to be the instrument for a remote She had a history of alcoholism, but had been sober for more that thirteen years. With this history, I expected to find some entity still attached to her. She mentioned that she had been feeling some anxiety while she was observing the releasing of other entities earlier in the day. We discussed whether we should deal with her anxiety first or at the end of the session. We decided to do the depossessions first and then deal with any anxiety that remained later.

The first remote scanning was for Henry K....age three in Boston, who had been ill and in pain. Dee's Inner Wisdom was consulted, approval was given, Henry was located at his home and permission from his high self was given in a different manner. The information transmitted was that he had a desire to be whole and out of pain and that he had a Light awareness.

Dee scanned Henry and found dark areas between his shoulders, from the back of his neck to the top of his head and down through his body to his genitals.

Dr. H. You who are causing the darkness in Henry's body come forward and identify yourself (of course there was the instruction to communicate through Dee without harming her in any way.) Do you have a name?

Entity Joe.

Dr. H. Joe, tell me about yourself. What happened to your body?

Joe I died many times.

Dr. H. What happened to the last body that you wore?

Joe I wasn't expecting it. It got dark.

Dr. H. What attracted you to Henry?

Joe I love Henry.

Dr. H. Did you know Henry before you died?

Joe I know him now.

Dr. H. Do you realize that you are having a detrimental effect on Henry?

Joe Henry doesn't seem full.

Dr. H. Does that mean that you are trying to fill him?

Joe Henry doesn't have enough energy for me.

Dr. H. Perhaps it's time that you left Henry and moved on. Have you considered that?

Joe I feel lost.

Dr. H. Has anyone told you about "The Light?"

Joe I know about "The Light."

Dr. H. It's the proper place to go when your body dies. Can you give me any good reason for not going there?

Joe I can't find one.

Dr. H. I call on anyone in "The Light" who cares about Joe to come for him, communicate with him and help him with his transition. Joe, look up. Tell me what you see.

Joe	Lots of color. It's pretty.

Dr. H.	Would you like to go there?

Joe	I'll think about it.

Dr. H.	I'll make a deal. You can go for a visit and if you don't like it you can come back---is that all right?

Joe	It's all right. Will it hurt?

Dr. H.	I ask those who came for Joe to show themselves to him. What do you see now Joe?

Joe	Beautiful people.

Dr. H.	Ask them what it's like there. What do they say?

Joe	They say that you can dance in "The Light" and you can feel the colors.

Dr. H.	That sounds like a pretty good place to go. Are you interested in going, Joe?

Joe	I sure am.

Dr. H.	Good. But before you go, look all around Henry's body and inside. Are there any others of your nature?

Joe	I don't want to leave Henry. Will he have enough life without me?

Dr. H.	We will take care of that. Your being there is actually harmful to Henry. I know that you intend to help, but that is not the effect you are having.

Joe	Am I degrading him?

Dr. H.	Look at yourself. Are you?

Joe	I'm pretty grey.

Dr. H.	As you leave Henry, perhaps you will take the greyness with you and he can be well. Is that all right with you?

Joe	I'll miss Henry.

Dr. H.	Are you willing to do this so he can be well?

Joe	I'm not sure.

Dr. H.	Think about it. You don't want Henry to stay ill, do you?

Joe	No.

Dr. H.	But as long as you are there you are keeping him ill. Do you realize that?

Joe	Yes. Let me feel "The Light" and see Henry without my greyness.

Dr. H.	You can come and visit Henry, but not be in his body. Is that all right with you?

Joe	Will I have the colored Light?

Dr. H.	Ask the beautiful people who came for you. What do they say?

Joe	Yes, and it's a lot of fun.

Dr. H.	Are there other beings in Henry that you can take with you?

Joe There's one deep in the heart. It's very old. I don't know this one.

Dr. H. Can you persuade the one deep in the heart to go with you?

Joe (no response)

Dr. H. Communicate with the one deep in the heart---see if that one will go with you. If not, we will deal with him later. But I would like your help.

Joe Will you let me go to "The Light" first?

Dr. H. That's no problem. Are there any others besides the one in Henry's heart?

Joe Yes, the one from the back of his shoulders up the back of his neck to the top of his head.

Dr. H. Is that one of your nature?

Joe More so, yes.

Dr. H. Then persuade that one to go with you.

Joe If I can be in "The Light", like my friends, then I'll reach down from "The Light" and get them.

Dr. H. Then go up and experience "The Light" a short while and then reach down for them. Tell me what happens.

Joe Ohhhh!

Dr. H. Are you getting the others?

Joe I just want to get to "The Light."

Dr. H. Keep talking to the others to persuade them to go with you. Are they listening?

Joe I think I can bring them with the head one. They see me in "The Light." They're curious and fascinated. The one in the heart, I don't know about him.

Dr. H. Take the ones that will go with you. Go in peace with our love and blessing. I ask the healing angels to fill the vacated spaces with healing Light.

Now, you located in Henry's heart, come forward and identify yourself. Do you have a name?

Entity No.

Dr. H. Did you ever have a name?

Entity Go away.

Dr. H. Have you ever had your own physical body?

Entity Go away. (The Forces of St. Michael were called to bring the net.)

Dr. H. Did you ever have your own human physical body?

Entity (Shouting) NO!

Dr. H. What is your purpose here in Henry?

Entity I'm not talking.

Dr. H. The net around you is growing tighter and tighter until you answer my questions promptly and truthfully. Have you ever had your own human physical body?

Entity I want Henry's sweetness.

Dr. H. Answer my question. Have you ever had your own human physical body?

Entity Not that I can remember.

Dr. H. What is your purpose for being in Henry, besides just wanting his sweetness?

Entity Go away.

Dr. H. What is your purpose?

Entity You go away.

Dr. H. I'm not going away. The net gets tighter and tighter until you answer my questions. What is your purpose?

Entity It's too tight. Get me out of here.

Dr. H. I want truthful answers. What is your purpose here?

Entity To darken this being's heart, physically, emotionally and spiritually.

Dr. H. Who sent you to Henry?

Entity I don't know.

Dr. H. Do you have a commander, a boss, who gives you your orders?

Entity I don't know.

Dr. H. What were your instructions when you came to Henry?

Entity I don't know.

Dr. H. Answer my questions truthfully. What were your instructions?

Entity I don't like you.

Dr. H. I'm not here to be liked. What were your instructions?

Entity I don't know. I told you before. I'm in his heart and I'm dark.

Dr. H. Were you told anything about the Light?

Entity What Light?

Dr. H. You seem to doubt that there is anything called Light. Is that correct?

Entity If there is Light, I will have no self.

Dr. H. This was part of your instructions, right?

Entity Maybe.

Dr. H. I will show you that your information is incorrect. I ask the Forces of St. Michael to send a shaft of light right through this "dark" one in Henry's heart, and another and another. That is Light. How does it feel?

Entity Take me. I want "The Light."

Dr. H. As you prepare for "The Light" there is something you must do. Look deep within the center of your own being and tell me what you find there.

Entity Anger, hate, dark.

Dr. H. Look beyond the anger, hate and dark. What do you find at your very center?

Entity There's a little light in me.

Dr. H. Watch that little light in you and tell me what happens as you watch it.

Entity Will I die?

Dr. H. Take the chance. I dare you!

Entity (Laughing) I'm not going to die.

Dr. H. What's really happening?

Entity The light's getting bigger and it takes away pain.

Dr. H. And you thought you wanted me to go away. Do you still want me to go away?

Entity You're not so bad.

Dr. H. Now that the Light is transforming you, is your darkness all gone?

Entity My darkness is gone. The little boy has a tender heart that is his own.

Dr. H. Is it dark?

Entity No.

Dr. H. Your being there has caused trouble, did you know that?

Entity (Nods) Do I get to stay in "The Light?"

Dr. H. Ask the beings around you. Some of them were dark like you were before you found your Light inside. Ask them what it's like where they exist.

Entity They say "The Light" gets stronger.

Dr. H. Do you like that?

Entity Yep!

Dr. H. Prepare to go with them, but before you leave, look all around Henry's body both inside and around the periphery for any other beings of your former dark nature that you can persuade to go with you. Are there any others?

Entity Not like me.

Dr. H. Fine. Prepare to go, and as you leave Henry's body call out through the un verse for all beings of your former dark nature---of equal rank and stature, of lesser rank and stature and greater rank and stature. Call to them to come and go home with you. Are you calling?

Entity No.

Dr. H. You get extra points---merit badges---if you do this. I'll ask again for you to call out through the universe for other beings to take with you. Are you calling.

Entity You're still here giving these instructions.

Dr. H. You said I wasn't so bad about giving instructions. Will you try it?

Entity All right. (Deep sigh)

Dr. H. Call out through the universe to all beings of your former dark nature to come and join you and go home with the rescuers. Are you calling?

Entity I don't want to call them. Can I shoot them with that little bit of Light?

Dr. H. You have to call them close first. Then maybe you can shoot them with Light. Then as they come ask each one to look inside and find their true nature at their core.

Entity That sounds good.

Dr. H. How many are coming?

Entity Dozens

Dr. H. Take them all with you and go in peace with our love and blessing. I now call the healing angels to fill with Light all the spaces where it is needed.

We then asked Dee to fill Henry's body with Light and let it expand to form a large bubble of Light around him, and bring her bit of consciousness back from Boston to her body.

Then Dee added a few comments about Henry: "The boy needs to be surrounded with Light often. He needs to be surrounded by those who understand him in "The Light." He needs his own guidance field to give him a lot of Light and he needs his own gentle space in the world."

When we told Dee that some of the experts had predicted that the child had not long to live, and asked her "Inner Wisdom" for response. she continued, "It will depend on whether the Light in him strengthens enough in the next few years. Then he will actually come to a different path and a different vibration than the one he is set on now. The path he is on now, he would gleefully finish in a few years, but he may choose to strengthen his life, to be here for those that he loves. There are those possibilities and we don't know yet which one will be."

We then inquired of Dee's Inner Wisdom---is it appropriate for Dee to scan Richard S. in Palo Alto? Approval was given. Dee was able to locate Richard and receive his high self's permission to scan him for the possible presence of attached entities.

Dee located dark areas in Richard's navel, lower internal organs and genitals. I asked the entity to come forward and indentify itself. No response. Then Dee said," It's a little girl. She's not good at language. She had a great desire to be Richard's child and stayed in his reproductive organs hoping Richard would give her the seed of life.

Dr. H. Little girl, do you know about "The Light"---a place to rest and learn while waiting for a new body?

Girl What is "The Light?"

Dr. H. It's a wonderful place of preparation for a new body. You will be doing much better by going to this place rather than staying here in a body which is not yours.

Girl Can I have a mom?

Dr. H. I call now on anyone in "The Light" who cares about this little girl, in the body of Richard, to come for her, greet her, communicate with her, and help her move to "The Light."

Girl Yes! Mom! (She left at once.)

Dr. H. I call now for the healing angels to bring Light to fill the space where the little girl was. (Dee began to shake her head.) What is it?

Dee The little girl wants us to tell Richard that she is O.K., that he doesn't have to wait for her or work for her or hope for her at any level, and he is not to worry. I agreed that we would let him know. Then Dee began to fill Richards body with Light, shaking her head all the while.

Dr. H. What is it?

Dee He needs a different diet that is clean and clear and not contaminated, that has not had a lot of processing and handled by people who don't care.

Dr. H. Has the Light filled Richard completely?

Dee Yes and extends around him in a bubble.

Dee brought her bit of consciousness back from Richard. We asked her Inner Wisdom if it would be appropriate for Dee to scan one more person.

Dee Can I have the whole Name?

Dr. H. It is K...M...H... in San Jose. (Dee shakes her head.) I then ask, is it appropriate to scan B...H....age 13,

Dee Yes.

Dr. H. Then send a bit of your consciousness to .... in San Jose, California and locate B...H..... .

Dee Can I have his middle name?

Dr. H. It's Kieth. (B... Kieth H....was located. Permission to scan him was granted by his high self.) Begin scanning and tell me what you find.

Dee His neck, the ends of his fingers and toes and bottoms of his feet.

The entity was called and asked for a name.

Entity Green.

Dr. H. Green? Have you had your own physical body?

Green Yes.

Dr. H. What happened to that body, Green?

Green I died in the hospital.

Dr. H. What was the cause of death?

Green Lungs.

Dr. H. What attracted you to this young man, B.... ?

Green He was easy.

Dr. H. How old was B... when you joined him?

Green There's someone else here. We joined at twelve and nine.

Dr. H. The one that I am speaking to---what is your name?

Green Elijah and Elizabeth.

Dr. H. Which is the spokesperson?

Green Elijah.

Dr. H. Elijah, tell me a little more about yourself. What happened when your body died?

Green It just slipped away.

Dr. H. Did anyone tell you about "The Light?"

Green It just slipped away.

Dr. H. What happened to you when your body slipped away?

Green How could it just slip away?

Dr. H. Go through the experience of having your body die. Tell me, what is happening to you as your body dies?

Green I'm in the hospital, and I get very light and my body is not with me any more.

Dr. H. Where are you when your body is not with you any more?

Green I don't know. I wish I knew.

Dr. H. Just examine your circumstances and you'll understand. Your body has just died.

Green It's a nice boy, a very nice boy.

Dr. H. It appears that you have decided to stay with this nice boy.

Green I don't know where else to go.

Dr. H. Had anyone told you about "The Light?"

Green I've seen "The Light.' It's high.

Dr. H. Was there any reason that you didn't go to "The Light?"

Green It's too far away.

Dr. H. I call now on anyone in "The Light" who cares about Elijah Green to come for him, greet him, help him to understand what is really happening. Elijah, look up. Tell me what you see.

Green It's getting closer. Could you bring it real close?

Dr. H. Sure. I ask those in "The Light" who care about Elijah to bring "The Light" closer so he will know you are really with him.

Green May I tell my sister?

Dr. H. Yes, and also tell anyone else of your nature that is in or around this young man. Are there others that you can persuade to go with you?

Green Just us two.

Dr. H. Then go in peace with our love and blessing, joining your loved ones who have come for you.

Green The boy is very nice. He is a very nice being.

Dr. H. You may thank him. Go now Elijah and Elizabeth. (We called the healing angels to fill the empty spaces and asked Dee to scan B....again. Dee then filled B....'s body with Light and expanded it around him. She was then asked to bring that bit of her consciousness back to her body and fill herself with Light.)

Dr. H. Tell me if you notice any areas of blockage or obstruction as your body fills with Light.

Dee From my neck down to my knees.

Dr. H. You who are causing the blockage in Dee's body, come forward, speak through her voice without harming her in any way and identify yourself. Do you have a name?

Entity (Shouting.) Ahhhhhhhh-hah!!

Dr. H. I call the forces of St. Michael to come and surround this one with a sturdy net of Light and pull it snug. I also ask that it be tightened slowly, but steadily until we get prompt and truthful answers from this one. Have you ever had a human physical body?

Entity **NO !!!**

Dr. H. You don't need to shout quite so loud. We can hear. What is your purpose here in Dee? (No response) I ask that the net be made tighter. What is your purpose?

Entity I just need a body.

Dr. H. How is it that you know that you need a body?

Entity (Whispering) Hang on. Hang on.

Dr. H. Who sent you to Dee?

Entity I found her.

Dr. H. What is your purpose?

Entity I hang on.

Dr. H. What effect are you having on Dee?

Entity Who cares?

Dr. H. I asked the question. I want prompt and truthful answers to my question. What is your purpose?

Entity BITCH!!

Dr. H. The net gets tighter and tighter. You must answer. What is your purpose?

Entity To take her life force.

Dr. H. How old was Dee when you joined her?

Entity Little.

Dr. H. Who sent you to Dee?

Entity I need her Light. I need her energy.

Dr. H. What were your instructions when you were sent to Dee?

Entity I don't remember instructions. I only remember the need. I don't know where the need came from. I NEED THE LIFE FORCE.

Dr. H. I'm asking you to review the procedure and the instructions before you joined Dee. As you review them, you will know who sent you and what the instructions were. Now what were the instructions?

Entity Can you help me? (This was a sudden change of tone apparently from the entity's realizing that it was in trouble.)

Dr. H. That is my purpose.

Entity How?

Dr. H. Answer my questions and I'll show you. What is your purpose?

Entity How do I know that you'll help me? (Much louder) I want a body! I want the life! I want the warmth and she has these.

Dr. H. Have you been observing what we have been doing here today?

Entity NO SHIT!

Dr. H. Have you observed that others of your nature have been helped?

Entity They are small.

Dr. H. Have you observed that they have been helped?

Entity (Softly) Maybe.

Dr. H. Look again and answer my question truthfully. Have you observed that they have been helped?

Entity (Shouting) YEH!!

Dr. H. I'm being difficult, I know.

Entity I saw them, BITCH!

Dr. H. Do you admit that they were helped?

Entity It looked like it.

Dr. H. Do you want help if I can do it?

Entity Yes.

Dr. H. Become aware of those beings around you outside the net. What do they look like?

Entity Whole.

Dr. H. Look again and tell me what they look like.

Entity If I go into the Lght, I will cease to exist.

Dr. H. You were led to believe that the Light would destroy you. Is that right?

Entity No more. Poof!

Dr. H. I ask the Forces of St. Michael to show this one by sending shafts of Light right through him. Another shaft and another and another. Did it destroy?

Entity Send some more.

Dr. H. I ask that more shafts of Light be sent through this dark one. Have they destroyed?

Entity More.

Dr. H. As many as you wish. I'm asking the Forces of St. Michael to continue. It felt pretty good, didn't it?

Entity More.

Dr. H. Don't be greedy. We have work to do. Those who led you to believe that Light would destroy, deceived and lied to you. How do you like associating with deceivers and liars? (Shakes head.) You don't have to any more. You can go with those who have come for you. The only alternative, since you have been found is to go to that awful place, the place of punishment. Have you been there?

Entity (Snarling.) Yeh!

Dr. H. What's it like, that awful place?

Entity Worse than what I'm doing with Dee right now.

Dr. H. Now that you have been found, there are only two choices open to you---to go to that awful place or to go with the beings that came for you to take you home to a wonderful place. It appears to me that you are about ready to go with them. Is that right?

Entity I want the Light.

Dr. H. Then look inside yourself to the very center. Tell me what you find at your center.

Entity My center is just Light.

Dr. H. Focus on that Light. What happens as you watch it?

Entity I'm just going to the Light. I want the Light.

Dr. H. You must get ready first. Let the Light within you fill you until it replaces all of your blackness. Tell me when that has happened.

Entity All gone.

Dr. H. I think you like this don't you?

Entity Yeh.

Dr. H. Now I have a job for you. I want you to look inside Dee and all around the periphery. Are there any other beings of your former dark nature?

Entity No.

Dr. H. As you leave Dee's body, call out through the universe to all beings of your former dark nature to come and go home with you and the rescuers. Are you calling? (nods) Are they coming.?

Entity Can I come back for them later?

Dr. H. It is part of your job to take them with you. You are earning extra bonus points by doing it this way. Are you calling?

Entity Yeh.

Dr. H. How many are coming?

Entity Three.

Dr. H. Call again.

Entity Twelve, eighteen.

Dr. H. Keep calling. One of your power should call many.

Entity Can they all come into the Light?

Dr. H. Sure. Tell them each to look inside themselves.

Entity Now lots of them are coming.

Dr. H. Keep calling. You are a powerful rescuer already and I thank you. Go in peace with our love and blessing.

I called the healing angels to come fill the vacated spaces and asked Dee to again fill her body with light. This time there were no further obstructions. Then after the usual post hypnotic suggestions, I helped Dee to waken. She reported that she felt much more free and comfortable.

This entity was one of the more difficult ones, but far from being dangerous or really threatening. I felt that it was just extra bratty, and that this quality was more a smokescreen for its fear than evidence that it could or would harm. The readiness with which it moved to the Light, illustrates that these "dark" beings are really dissatisfied with their state of being and their commanders and welcome a chance to escape.

This case, sent to me by Scott Herner of Scandia, Minnesota, concerns another young girl who for four years manifested both behavior and health problems. The remote depossession was requested by her mother. Mother reported that before the sudden change, Erin had been very pleasant, bright, happy, intelligent, always getting along well with her sisters. She had at about five, began to talk of having been abused, but could recall no details of actual abuse. At age seven, Erin developed headaches located from the top of her head and down the left side..

The changed pattern of behavior included vehement non-cooperation, obnoxiousness, being annoying, and mean to her younger sister--even to hitting her frequently.

Scott, with the help of an associate, Dodo S., located three Earthbound entities in Erin. The one in the head appeared to have tentacles going down the left side of her head. This one was Jim. Jim had died in the hospital and had joined Erin when she was being treated in the same hospital for a broken leg. Jim's presence was causing the headaches and her abhorrent behavior. Jim's father came for him in "The Light" but before Jim left he identified Agnes, in Erin's shoulder. Agnes had fallen outside on a cold day and had died before anyone found her. She had joined Erin for lack of a better place to go. Jim also named the one in Erin's left leg. This was Joe who refused to go to "The Light" with him, Agnes and Agnes' sister, Lenore, who had come for her.

After Jim left with his father, Agnes and Lenore, Scott dialogued with Joe. Joe petulantly stated that he "wasn't going anywhere." Further questioning brought the response that he had been very badly abused even while very young and had been beaten to death when he was four. He commented that he didn't think it was fair to die at such a young age. He had joined Erin when she was about five. Scott told Joe about "The Light" and that he could have his own body there. When the call was made for anyone in "The Light" to come for Joe, Grandma Annie came. Then Joe was willing to go.

Erin's body was filled and surrounded with Light. The very next morning, Erin was a changed child. She even told her mother that she felt different. She is much happier and quieter. This report was sent about five weeks after the depossession. During these five weeks the new attitude and behavior had continued. She gets along well with her sisters with no more hitting. Her aunt also remarked about the marked change in Erin.

The most significant part of this, I feel, is that Joe, who had been abused and beaten to death at four, had joined Erin when she was five. Erin began at that time to talk of being abused herself without being able to come up with any pertinent details. This raises the question as to whether other children may have felt abused because of an attached entity rather that from their own personal abuse. Could this be the reason for what has been called false memory syndrome?

THE COMPLETE HYPNOTHERAPIST

Most of us begin in hypnosis and hypnotherapy by learning first how to induce the hypnotic state, how to recognize the degree of relaxation and how to give direct and post-hypnotic suggestions. This is very important and the ideal starting place.

Direct suggestions have been useful in pain control, habit changes, sports performance, confidence, increased learning and in many other areas. For me, direct suggestions have never been enough. My intense curiosity led me to explore the mind to find its potentials. I did this exploring by asking questions of the hypnotized subjects. It soon became obvious that somewhere within us was a source of greater wisdom than we ordinarily exhibit. I chose to call this source the "Inner Wisdom"

I soon came to use this non-directive technique almost exclusively. This is a bit risky, as there is no way of determining what the answer will be beforehand. This uncertainty requires a high level of flexibility on the part of the hypnotherapist. Answers can't be anticipated with any degree of accuracy. When subjects are asked to go back in time and find the origin of a problem, the distance they travel can be a surprise. Some will find the cause of a problem or illness in a traumatic experience in childhood, or they may find it in a distant past time long before the birth of the present body. Without some understanding of our true nature, and the fact that we are eternal, have had earlier lives, and that events from previous lives impact on the present life, we will be completely lost when one of our hypnotized subjects tells us this sort of information.

In addition, we must continue to be aware that some problems are precipitated by the presence of attached entities, either Earthbound or "dark." We can only be the complete hypnotherapist by also developing knowledge and skills required to deal with entities, releasing them and sending them to the proper place.

It is essential also to realize that attached entities may not be the only contributing factor to a person's illness or problems. There are times when ongoing therapy is needed after all entities have been removed. If such therapy is needed, then we must realize it and refer them to whatever professional help is indicated if we are not qualified to care for the situation ourselves.

Sonnet

What spurs my ceaseless battle
That my world may better be?
What bids me on to conquer?
What matters life to me?

I wait not for the masses
To share my radiant dream.
I cannot wait 'til others know
The world can be redeemed.

I cannot wait 'til others share
This drive to right the wrongs --
Outgrowths of selfish greed and lust;
This battle, fight I must.

I cannot, must not, count the cost,
Be the battle won or lost.

by Irene Hickman, D. O.
as published in Alura
Poetry Journal, 1989

ADDENDUM

APPENDIX A

CONTRIBUTORS OF CASE REPORTS

Terry Ash (304) 375-5944
77 Greenwood Drive
Williamstown, West Virginia 26187

Rhonda Newhart (304) 422-3540
815 1/2 18th Street
Parkersburg, West Virginia 26101

Charlene Ackerman (608) 752-6744 or 757-0716
Olde Town Mall
20 South Main Street #29
Janesville, Wisconsin 53545

Olga Ivisn, M. D. (312) 878-2971
3322 South Calumet Street
Chicago, Illinois 60610

Ivan D. Goochey (402) 788-2579
Rt. 1 Box 56
Bennet, Nebraska 69317

Brian Knight, D. O. (512) 882-1001
1001 Santa Fe
Corpus Christi, Texas 78404

Linda Cox (313) 628-9424 After 8-94, area code will be (810)
2400 Farm Brook Trail
Oxford, Michigan 48370

Henry Smith-Rohrberg, Ph.D. (508) 548-3991
P. O. Box 344
West Falmouth, Massachusetts 02574

Scott Herner (612) 433 5373
12970 236th Street
Scandia, Minnesota 55073

APPENDIX B

WHAT IS A HUMAN BEING ?

Frederick Richards

Man has from the beginning, sought to know who, or rather what, he is. Man has sought through books, ideas, sages and through other men to find himself. Rarely does man search for knowledge about himself from the true source---*within.*

Man is a microcosm of the trinity, a sacred union of the temple, the scripture and the faith. These three components of man fit together like the pieces of a jigsaw puzzle, forming a complete whole that is part of God, Himself.

Man's temple is the body, a marvelous housing for the rest of his parts. The complexity of his own body is a mystery that man may never be able to understand with his intellect. So that man may learn during his time on Earth, the body provides five senses and the ability to form mental impulses through emotions. The body is also a mirror of the mind and spirit, reflecting in its own state of health, the health of the other two parts. However, the body, like a temple, ages and eventually returns to the dust from where it came.

The mind is interfaced with the body in an organ called the brain. The mind is not just a mass of tissue, but also a part of the spirit. The mind records information as memory, solves problems, dreams and creates new ideas. The mind is where all good and evil originate. The mind thinks and the body and the spirit react to this thought. As it is written, it shall be---and thus we create our own Heaven and Hell. This is because man is made in God's image and has free will. Man chooses to be capable or incompetent. And regardless of what his neighbor does, man chooses to love or hate him. Through man's free will he can choose good or evil. When he chooses evil and the root of evil---fear, he binds himself with the chains of fear and despair. All he really need fear is himself.

Like an aged book, when man dies, his thoughts and personality die with the body, filed on a forgotten shelf somewhere in the spirit.

The immaterial and often neglected part of man is the spirit. The spirit is connected to the body through seven "points" or chakras which are lined up along the spine. It is through these points that energies which keep man's body functioning, flow. These energies produced by the spirit not only keep man's body alive, but aid in healing and with the help of emotions produce spiritual and psychic phenomena. The spirit contains the book of life and records all thought and also has the ability to turn these thoughts into emotions. Emotions can block the flow of energy along the chakras as well as promote the flow, thereby producing physical and psychological disorder or order. The spirit is also connected to the brain and can be perceived as a voice of God. This part of the spirit attempts to guide man to correct thought and action in order to make man and his neighbors happy. The spirit gives proper guidance because it contains all the knowledge of the universe including knowledge of man's lives before life in the present body and knowledge of the experiences which have shaped him into his present state.

All knowledge in the universe is aççessible to the spirit through the spirit's contact with God. The consequences of man's actions are called karma. This most directly affects the spirit and teaches man so that eventually he might attain perfection.

Through karma, man really does reap what he sows, and if man really knew the consequences of his actions, perhaps his choices would be different. The spirit is man's hope, for when the body and conscious mind wither away, like a true and sacred faith, the spirit lives on.

The ancient Chinese believed that the root of all wisdom was to know one's self. If man would truly look inside himself and contact the God within, he could find the answers to the questions which have plagued his existence. Man could overcome despair and fear and find true peace.

If man would look *within,* he could eventually realize the hope of perfection.

(This was a prize-winning essay, written by a second-year student at Northeast Missouri State University in the early eighties. Frederick Richards is a pseudonym.)

"Knowledge without the
practical ability to apply same
may become sin."

Edgar Cayce

APPENDIX C

SUGGESTED READINGS.

William Baldwin, Ph.D.

Spirit Realeasement Therepy, a Techniques Manual.

Edith Fiore, Ph.D.

The Unquiet Dead.

Irene Hickman, D.O.

Mind Probe----Hypnosis.

(These three are the most important for those who plan to do remote depossession.)

Phyllis Atwater

Coming Back To Life.

Billy Joy Branson

Mining the Silver Lining.

Gina Cerminara

Many Mansions.

Anabel Chaplin

The Bright Light of Death.

Charles Cooke and A. E. Van Vogt

The Hypnotism Handbook.

Winifred B. Lucas, Ph.D.

Regression Therapy, a Handbook for Professionals 2 vol.

Raymond Moody, M.D.

Coming Back

Melvin Morse, M. D. and Paul Perry

Closer to the Light.

Thomas Sugrue

There is a River, a Biography of Edgar Cayce.

Brian L. Weiss, M. D.

Through Time into Healing.